AF251818

DIFFERENTIABILITY AND FRACTALITY IN DYNAMICS OF PHYSICAL SYSTEMS

DIFFERENTIABILITY AND FRACTALITY IN DYNAMICS OF PHYSICAL SYSTEMS

Ioan Merches

Alexandru Ioan Cuza University, Iasi, Romania

Maricel Agop

Gheorghe Asachi Technical University, Iasi, Romania

NEW JERSEY · LONDON · SINGAPORE · BEIJING · SHANGHAI · HONG KONG · TAIPEI · CHENNAI · TOKYO

Published by

World Scientific Publishing Co. Pte. Ltd.

5 Toh Tuck Link, Singapore 596224

USA office: 27 Warren Street, Suite 401-402, Hackensack, NJ 07601

UK office: 57 Shelton Street, Covent Garden, London WC2H 9HE

Library of Congress Cataloging-in-Publication Data

Merches, Ioan, author.

   Differentiability and fractality in dynamics of physical systems / Ioan Merches (Alexandru Ioan Cuza University, Romania), Maricel Agop (Gheorghe Asachi Technical University, Romania).

     pages cm

   Includes bibliographical references and index.

   ISBN 978-9814678384 (hard cover : alk. paper)

   1. Differentiable dynamical systems. 2. Fractals. 3. Mathematical physics. I. Agop, Maricel, author. II. Title.

   QA614.8.M46 2015

   515'.39--dc23

                       2015006096

British Library Cataloguing-in-Publication Data

A catalogue record for this book is available from the British Library.

In-house Editor: Christopher Teo

Typeset by Stallion Press

Email: enquiries@stallionpress.com

Printed by FuIsland Offset Printing (S) Pte Ltd Singapore

*The true sign of intelligence
is not knowledge but imagination.*

Albert Einstein

Preface

This monograph is a collection of ideas developed by several authors on differentiability and fractality, taken as basic frames of approaching the study of physical systems dynamics. Here are some results we want to put into evidence, together with the specific "arsenal" of methods and procedures.

Using the 1-forms in Cartan sense theory, we are offering an elegant procedure of analyzing the motion, in order to obtain the main results of the theory of relativity, and some new results as well. This way, the form of the inertia transformations — *i.e.* the geometric structure of the space-time "aggregate" — is not at all prejudiced. In other words, the irreducible attributes of motion of a body, two bodies, or a solid rigid are dependent on a set of Euclidean invariants. Within such an approach, without abandoning the space-time standard frame, gravitation is re-found as an inertial mass interaction. In addition, a repulsive interaction — responsible for Hubble's dilatation — is found; this interaction is very sensitive at intergalactic distances.

The wave-corpuscle duality, reducible to the equivalence of two 1-forms, inertial and ondulator, imply motions on curves of constant informational energy. This way not only quantization is statuated, but also dynamics implying velocity limits — not necessarily the speed of light in vacuum.

Motion of a charged particle in a combined field (composed by an electromagnetic and a magnetic constant fields) mimes by numeric simulation not only various scenarios of evolution towards chaos (period doubling, bifurcations on subharmonics, etc.), but also pattern generation (motion on stationary orbits, etc.). Under these circumstances, trajectories are continuous and non-differentiable curves (fractal curves), and this fact imposes a new space-time geometry, *i.e.* the fractal geometry.

There is also developed a dynamics of the physical object in a fractal space-time, together with some instructive and expressive applications: the fractal model of the atom, the potential gap, the linear harmonic oscillator, the free particle, Heisenberg's uncertainty relations, etc.

The physical systems that display chaotic behavior are recognized to acquire self-similarity and manifest strong fluctuations at all possible scales. Since fractality appears as an universal property of these systems, it is necessary to construct a new physics, either through non-differentiability (Nottale's scale relativity model), or through fractional derivative (fractional physics model). In the first variant, an extended model of Nottale's theory (Scale relativity theory with an arbitrary constant fractal dimension) is established and some applications (fractal model of the atom, particle in a box, harmonic oscillator and free particle) are given. In the second variant, the theory of fractional scale relativity and some applications (fractal covariant mechanics induced by modified scale laws, fractional Schrödinger equation, and emergence of complex gravity) are obtained and discussed.

At the end of this introductory presentation we want to emphasize the fact that our endeavor is addressed to researches, scientists and all those working in the field. Even if this book is not a textbook, it can successfully be used by MSc and PhD students interested in this unlimited and fascinating subject.

The authors express their gratitude to Professor Ieronim Mihaila, from the Faculty of Mathematics, University of Bucharest, for his competent advice during elaboration of the book.

* * *

We dedicate this monograph to the memory of Professor Octav Onicescu, prominent researcher and scientist, member of the Romanian Academy.

The Authors
Iasi, November 2014

Contents

Chapter 1

Principles of Motion in Invariantive Mechanics

1.1. The Euler-Lagrange and Hamilton's equations obtained by means of exterior forms

Let

$$L = L(q^1, q^2, \ldots, q^n, \dot{q}^1, \dot{q}^2, \ldots, \dot{q}^n, t) \equiv L(q, \dot{q}, t) \qquad (1.1)$$

be a function of class C^2, called *Lagrangian*, where $q \equiv (q^1, q^2, \ldots, q^n)$ are the generalized coordinates,

$$\dot{q}^j = \frac{dq^i}{dt} \quad (j = \overline{1,n}) \qquad (1.2)$$

are the generalized velocities, and t is the time. Let also S be the *action*, defined by the exterior 1-form

$$\omega = dS = L\,dt. \qquad (1.3)$$

Theorem 1.1. *The total derivative of the action with respect to time yields the Hamilton-Jacobi equation* [1]

$$\frac{\partial S}{\partial t} + H = 0. \qquad (1.4)$$

Indeed, if the functional dependence

$$S = S(q, t) \qquad (1.5)$$

as well as the definition of generalized momenta

$$p_j = \frac{\partial S}{\partial q^j} \quad (j = \overline{1,n}) \qquad (1.6)$$

1

are accepted, the total derivative of S with respect to time is

$$\frac{dS}{dt} = \frac{\partial S}{\partial t} + \sum_{j=1}^{n} \frac{\partial S}{\partial q^j} \dot{q}^j \overset{\text{def}}{=} \frac{\partial S}{\partial t} + p\dot{q}. \tag{1.7}$$

(N.B. Here and hereafter, the summation symbol and the summation indices shall be omitted).

Since, according to (1.3)

$$\frac{\partial S}{\partial t} = L, \tag{1.8}$$

relations (1.7) and (1.8) lead to (1.4), where the *Hamiltonian H* is defined as

$$H = p\dot{q} - L. \tag{1.9}$$

Theorem 1.2. *Closeness of the 1-form* (1.3)

$$D\omega = dL \wedge dt = 0 \tag{1.10}$$

reduces to Euler-Lagrange equations [1]

$$\frac{d}{dt}\left(\frac{\partial L}{\partial \dot{q}}\right) - \frac{\partial L}{\partial q} = 0 \tag{1.11}$$

and the integral of motion [1]

$$\frac{dH}{dt} = -\frac{\partial L}{\partial t}. \tag{1.12}$$

Written explicitly, relation (1.10) becomes

$$D\omega = dL\delta t - \delta L dt = 0. \tag{1.13}$$

According to (1.1), we can write

$$\delta L = \frac{\partial L}{\partial q}\delta q + \frac{\partial L}{\partial \dot{q}}\delta \dot{q} + \frac{\partial L}{\partial t}\delta t, \tag{1.14}$$

On the other hand, operators d and δ commute, so that

$$\frac{\partial L}{\partial \dot{q}}\delta \dot{q} = \frac{\partial L}{\partial q}\delta \frac{dq}{dt} = \frac{\partial L}{\partial q}\frac{d}{dt}\delta q. \tag{1.15}$$

Equation (1.3) can then be written as

$$\left[\frac{d}{dt}\left(\frac{\partial L}{\partial \dot{q}}\right) - \frac{\partial L}{\partial q}\right]\delta q + \left[\frac{d}{dt}\left(L - \frac{\partial L}{\partial \dot{q}}\dot{q}\right) - \frac{\partial L}{\partial t}\right]\delta t = 0. \qquad (1.16)$$

Using the Hamiltonian defined by (1.9) and the fact that variations δq and δt are arbitrary, equation (1.16) yields (1.11) and (1.12). If, in particular, L does not explicitly depend on time, then (1.12) reduces to the energy conservation law

$$\frac{dH}{dt} = 0. \qquad (1.17)$$

As one can see, the 1-form (1.3) can also be written as

$$\omega = [p\dot{q} - H]dt = pdq - Hdt, \qquad (1.18)$$

where the *generalized momentum p* is defined by

$$p = \frac{\partial L}{\partial \dot{q}}. \qquad (1.19)$$

Theorem 1.3. *Closeness of the 1-form* (1.18)

$$D\omega = dp \wedge dq - dH \wedge dt = 0 \qquad (1.20)$$

reduces to Hamilton's canonical equations [1]

$$\dot{q} = \frac{\partial H}{\partial p} \quad \dot{p} = -\frac{\partial H}{\partial q} \qquad (1.21)$$

together with the integral of motion [1]

$$\frac{dH}{dt} = \frac{\partial H}{\partial t}. \qquad (1.22)$$

Explicitating (1.20), we have

$$dp\delta q - \delta pdq - dH\delta t + \delta Hdt = 0. \qquad (1.23)$$

Since $H = H(p, q, t)$, one can write

$$\delta H = \frac{\partial H}{\partial q}\delta q + \frac{\partial H}{\partial p}\delta p + \frac{\partial H}{\partial t}\delta t, \qquad (1.24)$$

and (1.23), after some simple manipulation, becomes

$$\left(-\dot{q} + \frac{\partial H}{\partial p}\right)\delta p + \left(\dot{p} + \frac{\partial H}{\partial q}\right)\delta q + \left(-\frac{dH}{dt} + \frac{\partial H}{\partial t}\right)\delta t = 0. \quad (1.25)$$

Equating to zero the coefficients of variations δq, δp, and δt, equations (1.21) and (1.22) follow immediately. In particular, if H does not explicitly depend on time, (1.22) reduces to the energy conservation law (1.17).

1.2. The Cartan motion principle

The previous results allow us to operate with both Lagrange function L and Hamilton's function H. This means that, as long as the canonical formalism operates, Hamilton's canonical equations (1.16) and Euler-Lagrange equations (1.5) are equivalent. But, if the canonical formalism is not applicable, the two systems of equations are — in general -different. When passing from the Lagrangian to canonical formalism, not all canonical momenta given by (1.19) are defined [2]. Between the two functions, Lagrangian and Hamiltonian, the invariantive mechanics chooses the last one. Here are some arguments in this respect:

1. In general, the Lagrangian does no have a direct physical signification (except for the case when it can be separated in two terms, of kinetic and potential energy significance).
2. The Hamiltonian represents not only energy, but also generates the motion (see paragraph concerning the Onicescu informational energy). This function is conserved in both previously discussed alternatives.
3. Using the Hamiltonian H and the exterior forms, one can construct the following principle (*Cartan's principle*):

The law of motion for a discrete system with n components, characterized by the inertial 1-form

$$\omega = pdq - Hdt \qquad (1.26)$$

is given by the cancellation of the exterior derivative of 1-form (1.26), that is

$$D\omega = dp\delta q - \delta p dq - dH\delta t - \delta H dt = 0. \qquad (1.27)$$

The law of motion for a discrete system of n components, characterized by 1-form (1.26), is given by its closeness

$$D\omega = dp \wedge dq - dH \wedge dt = 0. \qquad (1.28)$$

Observation 1.1. Euler-Lagrange and Hamilton's canonical equations keep their form in case of systems with an infinite number of degrees of freedom (continuous deformable media and fields), except for the facts that L and H are replaced by their densities $\mathcal{L}$ and $\mathcal{H}$, while the usual derivatives are replaced by functional derivatives (see, for details, reference [4]).

Chapter 2

Inertial Invariantive Motion of the Material Point

2.1. Preliminaries

The exterior forms allow one to study the inertial motion of the particle (material point). In other words, using the inertial 1-form and the motion principle, one can obtain not only the conservation laws for momentum and energy, but also the structure relations for these fundamental quantities. Description of the motion can be performed either by taking $\epsilon = +1$, $\omega = c$, approaching the special relativity ([3], [4], [5], [6], [7]), or by choosing $\epsilon = -1$, leading to special physical realities (particles and antiparticles, quantum effects induced by the exotic matter, etc.),

2.2. Inertial motion of the particle (material point)

Postulate 2.1. *The space-time is structured as $E_3 \times T$ variety.*

Consequently:

i. Velocity $\mathbf{v}$ is defined by

$$\mathbf{v} = \frac{d\mathbf{r}}{dt},\tag{2.1}$$

   where t is the time.
ii. The inertial 1-form (1.18) can be written as

$$\omega^i = \mathbf{p}d\mathbf{r} - H\,dt\tag{2.2}$$

here $\mathbf{p}$ being momentum, and H — the Hamiltonian. The dot of the scalar product has been omitted.

7

iii. Isotropy and homogeneity of the space demand the functional dependence

$$H = H(\alpha), \tag{2.3}$$

where α is the Euclidean invariant

$$\alpha = \frac{1}{2}\mathbf{p}^2. \tag{2.4}$$

Postulate 2.2. *The inertial motion of the particle is described by the closeness of the inertial 1-form (2.2)*

$$D(\omega^i) = d\mathbf{p} \wedge d\mathbf{r} - dH \wedge dt = 0, \tag{2.5}$$

or, explicitly

$$D(\omega^i) = d\mathbf{p}\delta\mathbf{r} - \delta\mathbf{p}d\mathbf{r} - dH\delta t - \delta H dt = 0. \tag{2.5$'$}$$

Since according to (2.3) and (2.4) δH is

$$\delta H = \frac{\partial H}{\partial \alpha}\mathbf{p}\delta\mathbf{p}, \tag{2.6}$$

relation (2.5) becomes

$$D(\omega^i) = d\mathbf{p}\delta\mathbf{r} - dH\delta t + \left(-d\mathbf{r} + \frac{\partial H}{\partial \alpha}\mathbf{p}dt\right)\delta\mathbf{p} = 0. \tag{2.7}$$

Since variations $\delta\mathbf{r}, \delta t, \delta\mathbf{p}$ are arbitrary, relation (2.7) yields:

i. The momentum conservation law

$$\frac{d\mathbf{p}}{dt} = 0; \tag{2.8}$$

ii. The energy conservation law

$$\frac{dH}{dt} = 0; \tag{2.9}$$

iii. The momentum definition relation

$$\mathbf{p} = m\mathbf{v}, \tag{2.10}$$

with

$$m = \left(\frac{\partial H}{\partial \alpha}\right)^{-1}, \tag{2.11}$$

called *inertial mass.*

If the inertial mass is a constant ($m = m_0 = const.$), then (2.6), by means of (2.11), becomes

$$-\mathbf{p}\delta\mathbf{p} + m\delta H = 0. \tag{2.12}$$

Taking $H = 0$ for $\mathbf{p} = 0$, the classical mechanics result is obtained:

$$H = \frac{p^2}{2m_0}. \tag{2.13}$$

In the same contest, m_0 is called *rest mass*, and momentum writes $\mathbf{p} = m_0\mathbf{v}$. Since the material point has an irreductible simplicity, its inertial motion on space-time $E_3 \times T$ differs from that of another particle by mass (making abstraction of position and velocity). Therefore, the kinematic 1-form

$$\omega^c = \mathbf{v}d\mathbf{r} - Hdt, \tag{2.14}$$

obtained from (2.2) by means of $\omega^i = m\omega^c$, has to be the same for any particle. This observation implies the following postulate:

Postulate 2.3. *The ratio $\frac{H}{m}$ is a universal constant.*

Since the units of this ratio and of a squared velocity are the same, that is $\frac{H}{m} = \epsilon\omega^2$, with $\epsilon = \pm 1$, it follows that the Hamiltonian of a free particle writes

$$H = \epsilon m\omega^2. \tag{2.15}$$

In view of (2.5) and (2.15) we then have

$$\delta H = \epsilon\omega^2\delta m = \mathbf{v}\delta\mathbf{p}, \tag{2.16}$$

Multiplying by m and using (2.10), we still have

$$\epsilon\omega^2\delta m = \mathbf{p}\delta\mathbf{p}. \tag{2.17}$$

The last relation can also be written as

$$\delta(\mathbf{p}^2 - \epsilon\omega^2 m^2) = 0, \tag{2.18}$$

and, by integration

$$\mathbf{p}^2 - \epsilon\omega^2 m^2 = const. \tag{2.19}$$

To determine the constant of integration one takes $m = m_0$ for $\mathbf{p} = 0$, where m_0 is the rest mass. One finds

$$\mathbf{p}^2 - \epsilon\omega^2 m^2 = -\epsilon\omega^2 m_0^2, \tag{2.20}$$

In view of (2.10), the velocity dependence of the inertial mass of a free particle writes

$$m = m_0 \left(1 - \epsilon\frac{v^2}{\omega^2}\right)^{-\frac{1}{2}}. \tag{2.21}$$

Using (2.19), the momentum therefore is

$$\mathbf{p} = m_0 \mathbf{v} \left(1 - \epsilon\frac{v^2}{\omega^2}\right)^{-\frac{1}{2}}, \tag{2.22}$$

while (2.11) yields the energy

$$H = m_0 \epsilon\omega^2 \left(1 - \frac{v^2}{\omega^2}\right)^{-\frac{1}{2}}. \tag{2.23}$$

The mass m has a physical significance for $\epsilon = 1$ and $\left(1 - \frac{v^2}{\omega^2}\right)^{-\frac{1}{2}} > 0$, which implies $v < \omega$. In other words, ω stands for a limit velocity. From $\mathbf{p} = const.$, for $m \neq 0$, it follows the law of motion $\mathbf{v} = const.$, saying that the particle moves uniformly along a straight line. In other words, the inertia principle is satisfied. In invariantive mechanics, ω is an arbitrary speed limit, being an attribute of light only by extrapolation of electromagnetic phenomena to the other phenomena of the physical world.

The magnitude of ω is not apriorically restricted; in principle, it can be either smaller or higher then the speed of light, its value being decided only by the experiment. In particular, for $\omega \equiv c$, where

c is the speed of light in vacuum, this model leads to the well-known results of the special relativity:

$$m = m_0 \left(1 - \frac{v^2}{c^2}\right)^{-\frac{1}{2}},$$

$$\mathbf{p} = m_0\mathbf{v} \left(1 - \frac{v^2}{c^2}\right)^{-\frac{1}{2}}, \qquad (2.24)$$

$$H = E = mc^2 = c\sqrt{m_0^2 c^2 + \mathbf{p}^2}.$$

For $v \ll c$, (2.24) go to

$$m \approx m_0, \quad \mathbf{p} \approx m_0\mathbf{v}, \quad H \approx m_0 c^2 + \frac{1}{2}m_0 v^2,$$

where $m_0 c^2$ is the rest energy, and $\frac{1}{2}m_0 v^2$ the kinetic energy.

The choice $\epsilon = 1$, $\omega = c$, $H = mc^2$ in (2.20) leads to

$$H = \pm c\sqrt{m_0^2 c^2 + \mathbf{p}^2}.$$

A similar situation appears in quantum relativistic theory [4] connected to the two possible interpretations of the Dirac equation (electron-positron). Here, the positive solution for energy given by $H = c\sqrt{m_0^2 c^2 + \mathbf{p}^2}$ is associated with the electron, and negative solution $H = -c\sqrt{m_0^2 c^2 + \mathbf{p}^2}$ to the positron.

If in (2.21), (2.22), and (2.23), we choose $\epsilon = -1$, we obtain:

$$m = m_0 \left(1 + \frac{v^2}{\omega^2}\right)^{-\frac{1}{2}};$$

$$\mathbf{p} = m_0\mathbf{v} \left(1 + \frac{v^2}{\omega^2}\right)^{-\frac{1}{2}};$$

$$H = -m\omega^2 = -\omega\left(m_0^2 c^2 - \mathbf{p}^2\right)^{\frac{1}{2}}.$$

This time the mass is real, and there is no limit for velocity. The mass decreases when velocity increases, and according to

$$m = m_0 \left(m_0^2 - \frac{\mathbf{p}^2}{\omega^2}\right)^{\frac{1}{2}}$$

admits the rest mass as a superior limit

$$m \leq m_0.$$

This way, due to the unlimited increase of velocity, the inertia is susceptible to an unlimited decrease to zero; this situation corresponds to an infinitely small moving mass, that is to non-inertia. If the rest mass is effectively zero (such as field particles) condition for a mass smaller that the rest mass is out of question, so that mass cannot decrease with the motion. It depends on some other characteristics of the motion (such as frequency ν through relation $m = h\nu/\omega^2$, where h is the Planck constant). This fact presents a difficulty: the 'particle' cannot propagate, it constantly remains at rest. Therefore, such a 'particle' must have a non-zero rest mass, expressing an inertia which depends on the velocity ω.

The same alternative ($\epsilon = -1$), together with $H = -m\omega^2$, yield

$$H = \pm\omega\left(m_0^2\omega^2 - \mathbf{p}^2\right)^{\frac{1}{2}}.$$

Therefore, for $m_0\omega \geq |\mathbf{p}|$, there are both 'particles' associated with the positive energy $H = \omega(m_0^2\omega^2 - \mathbf{p}^2)^{\frac{1}{2}}$, and 'anti-particles' associated with negative energy $H = -\omega(m_0^2\omega^2 - \mathbf{p}^2)^{\frac{1}{2}}$. Physically, a negative energy would correspond to a deceleration process when a force is applied.

Recently, the negative energy states have been correlated either to Casimir quantum effects (interaction between two conductor plates under the action of electron-positron vacuum), Unruh radiation (acceleration of frames acted by the electron-positron vacuum), Hawking radiation (black body radiation that is predicted to be released by black holes, due to quantum effects near the event horizon) ([7], [8]), gravitational superconductivity (generation of Cooper pairs of gravitational type), or to some special topological space-time structures, capable to connect very far zones of the Universe by the so-called *wormholes*, conceived as extra-dimensional tunnels (with respect to the metagalactic hyperplane) [9]. These energy states imply a special kind of matter, called *exotic matter* [8], [9].

2.3. The Minkowski-Einstein universe. The Lorentz metric and Lorentz transformations

Writing (2.2) in the form

$$\omega^i = -m\frac{\omega^2 dt\delta t - d\mathbf{r}\delta\mathbf{r}}{dt},\tag{2.25}$$

it follows that the bilinear form

$$\phi = d\mathbf{r}\delta\mathbf{r} - \omega^2 dt\delta t\tag{2.26}$$

and the quadratic form

$$d\sigma^2 = c^2 dt^2 - d\mathbf{r}^2\tag{2.27}$$

are simultaneously invariant with respect to a continuous Lorentz-Poincaré 10-parameter group of linear transformations. Since

$$m = \frac{m_0 dt}{\sqrt{\omega^2 dt^2 - (d\mathbf{r})^2}} = \frac{m_0 dt}{d\sigma}\tag{2.28}$$

or, still

$$\frac{m}{dt} = \frac{m_0}{d\sigma},\tag{2.29}$$

it follows that, together with invariance of the 2-form $d\sigma^2$, that is of the bilinear form ϕ, the inertial 1-form

$$\omega^i = -m_0\frac{\omega^2 dt\delta t - d\mathbf{r}\delta\mathbf{r}}{d\sigma}$$

is invariant with respect to the same Lorentz-Poincaré type group.

Expression $\omega^2 dt\delta t - d\mathbf{r}\delta\mathbf{r}$ is a scalar quasi-product in the metric

$$d\sigma^2 = \omega^2 dt^2 - d\mathbf{r}d\mathbf{r}.\tag{2.30}$$

The 4-dimensional space $(\mathbf{r}, t)$ endowed with this metric is of Einstein type. Motions starting from the point $(\mathbf{r}, t)$ are real only if

$$d\sigma^2 > 0,\tag{2.31}$$

while motions corresponding to

$$d\sigma^2 = 0, \tag{2.32}$$

which are performed with velocity ω, are not fulfilled by material particles, but only by field particles. This way, velocity ω represents a velocity limit for material points.

The variety

$$\omega^2 \xi_0^2 - \xi_1^2 - \xi_2^2 - \xi_3^2 = 0, \tag{2.33}$$

where $(\xi_0, \xi_1, \xi_2, \xi_3)$ are the parameters of the direction of motion for which

$$\frac{dx_1}{\xi_1} = \frac{dx_2}{\xi_2} = \frac{dx_3}{\xi_3} = \frac{dt}{\xi_0}, \tag{2.34}$$

is a hypercone with its apex at the point of coordinates (x_1, x_2, x_3, t) of the 4-dimensional space $(\xi_0, \xi_1, \xi_2, \xi_3)$. It separates this space into two domains: the interior domain belongs to the universe of motions, being submitted to restriction

$$\omega^2 \xi_0^2 - \xi_1^2 - \xi_2^2 - \xi_3^2 > 0. \tag{2.35}$$

The moving point (x_1, x_2, x_3, t) carries with it the separation cone of the universe of motions.

The linear transformations which leave invariant the form $\omega^2 \xi_0^2 - \xi_1^2$, characterized by $\xi_2 = \xi_2^0$, $\xi_3 = \xi_3^0$ are Lorentz-type transformations. Their form is

$$\begin{cases} \xi_0' = \alpha \xi_0 + \beta \xi_1, \\ \xi_1' = \gamma \xi_0 + \delta \xi_1, \end{cases} \tag{2.36}$$

while the coefficients of the transformation verify the identity

$$\omega^2 \xi_0^2 - \xi_1^2 = \omega^2 \xi_0'^2 2 - \xi_1'^2. \tag{2.37}$$

Then there exists an argument φ for which

$$\begin{cases} \alpha = \cosh \varphi; & \beta = -\frac{1}{\omega} \sinh \varphi; \\ \gamma = -\omega \sinh \varphi; & \delta = \cosh \varphi. \end{cases} \tag{2.38}$$

Denoting

$$u = -\omega \tanh \varphi, \tag{2.39}$$

relations (2.36) yield the Lorentz-type transformations

$$\xi_0' = \frac{1}{\sqrt{1 - \frac{u^2}{\omega^2}}} \left(\xi_0 + \frac{u}{\omega^2} \xi_1 \right); \quad \xi_1' = \frac{1}{\sqrt{1 - \frac{u^2}{\omega^2}}} (\xi_1 + u\xi_0). \tag{2.40}$$

representing a continuous group with a parameter

$$\frac{\xi_1'}{\xi_0'} = \frac{\frac{\xi_1}{\xi_0} + u}{1 + \frac{1}{\omega^2} \frac{\xi_1}{\xi_0} u}. \tag{2.41}$$

Since the quantities ξ_0 and ξ_1 are proportional to dt and dx_1, respectively, the substitutions

$$\frac{\xi_1'}{\xi_0} = \frac{dx}{dt} = v; \quad \frac{\xi_1'}{\xi_0'} = \frac{dx'}{dt'} = v' \tag{2.42}$$

performed in (2.41) give the velocity composition law

$$v' = \frac{v + u}{1 + \frac{vu}{\omega^2}}. \tag{2.43}$$

According to this law, for $v \ll \omega$, $u \leq \omega$, the composite velocity also satisfies the inequality

$$v' \leq \omega.$$

The equality sign is valid only if either v or u equals ω. Indeed, (2.43) gives

$$\frac{v}{\omega} + \frac{u}{\omega} < 1 + \frac{vu}{\omega^2}$$

which is equivalent to the obvious inequality

$$\frac{v}{\omega} \left(1 - \frac{u}{\omega} \right) > 1 - \frac{u}{\omega}$$

with $v < \omega$ and $u < \omega$. For $v = \omega$ and $u = \omega$, the equality sign is obtained.

For $\omega = c$, (2.30) reduces to Minkowski metric, while (2.36) go to Lorentz transformations. In this context, the quantities u, v, v' have the usual significance of special relativity. If $u \ll \omega$, (2.43) reduces to Galilean law of velocity addition

$$v' = v + u. \tag{2.44}$$

2.4. The anti-Minkowski type Universe

Considering the alternative $\epsilon = -1$, one can see that the 1-form (2.2) written as

$$\omega^i = m \frac{d\mathbf{r}\delta\mathbf{r} + \omega^2 dt\delta t}{dt}$$

implies the bilinear form

$$\phi = d\mathbf{r}\delta\mathbf{r} + \omega^2 dt\delta t$$

which, together with the quadratic form

$$d\sigma = \omega^2 dt^2 + d\mathbf{r}d\mathbf{r}$$

are simultaneously invariant with respect to the 4-dimensional roto-translation group.

Since the anti-Minkowski type space has a structure of an Euclidean space, its fibers are geometrically much simpler than those of a Minkowski type space. They have neither motions of zero length, nor zones with forbidden displacements.

The fibres are homogeneous Euclidean universes, which are different from each other by values of ω. Consequently, the motion can be described for one and the same ω, that is within the same fiber.

2.5. Inertial motion of the particle in the Minkowski-Einstein type universe

Since the Lorentz-type transformations lead to a Minkowski-type space, the inertial motion of the particle can be expressed in a four-dimensional form. Indeed, using the inertial 1-from (2.2) organized as

$$\omega^i = \mathbf{p} \cdot d\mathbf{r} + \frac{i}{\omega} Ei\omega dt \tag{2.45}$$

one can construct the following four-vectors:

i. The momentum 4-vector p_β, with components

$$p_\beta = \left(\mathbf{p}, \frac{i}{\omega}E\right), \quad (\beta = \overline{1,4}); \tag{2.46}$$

ii. The position 4-vector dx_β given by

$$dx_\beta = (\mathbf{dr}, i\omega dt). \tag{2.47}$$

Postulate 2.4. *The inertial motion of the particle (material point) in a Minkowski-type universe is described by the closeness of the 1-form (2.45)*

$$D(\omega^i) = dp_\beta \wedge dx_\beta. \tag{2.48}$$

Explicitating (2.48), we have

$$D(\omega^i) = dp_\beta \delta x_\beta - \delta p_\beta dx_\beta = 0.$$

Multiplying this relation by $(d\tau)^{-1}$ and equating zero the coefficient of δx_β, it follows:

i. The law of conservation of the momentum 4-vector

$$\frac{dp_\beta}{d\tau} = 0; \tag{2.49}$$

ii. The structure relation

$$u_\beta \delta u_\beta = 0, \tag{2.50}$$

that is the 4-vectors u_β and δu_β are orthogonal.

Chapter 3

Field Invariantive Theories

3.1. Preliminaries

The external forms allow one to study the field behavior. More precisely, the use of 1-forms leads to an invariantive linear field theory, while the 2-forms lead to an invariantive nonlinear theory. Our case shall be developed choosing the alternative $\epsilon = -1$, $\omega = c$, in which case the motion is described by the Minkowski space-time. These results shall be then transposed for the anti-Minkowskian case $(\epsilon = -1)$.

As we shall see, the invariantive nonlinear theory reduced, for a particular case, to the nonlinear Born-Infeld electrodynamics.

Consequently, the invariantive field theory can represent a new way of approach of physical fields. It allows a unitary description of various fields and, in case of nonlinear fields, gives rise to the definition of a conservative energy tensor ([5], [6], [10], [11], [12], [13]).

3.2. Linear field theory

3.2.1. *Field invariantive equations*

To describe the linear field, let us introduce the 1-form ([14].[15])

$$\omega^p = \mathbf{A}d\mathbf{r} - Vdt = \mathbf{A}d\mathbf{r} + \frac{i}{c}Vicdt = A_\beta dx_\beta, \quad (\beta = \overline{1,4}), \quad (3.1)$$

where A_β is the potential 4-vector

$$A_\beta = \left(\mathbf{A}, \frac{i}{c}V\right), \quad (3.2)$$

19

with $\mathbf{A}$ the vector potential, V the scalar potential, and $\mathbf{r}$ the position vector in Euclidean space E_3. We shall call it the potential 1-form.

The exterior differential of the 1-form (3.1)

$$D(\omega^p) = dA_\beta \wedge dx_\beta = dA_\beta \delta x_\beta - \delta A_\beta dx_\beta$$

$$= (\partial_\sigma A_\beta - \partial_\beta A_\sigma)\partial x_\sigma \wedge dx_\beta \tag{3.3}$$

both the antisymmetric tensor

$$F_{\sigma\beta} = \partial_\sigma A_\beta - \partial_\beta A_\sigma, \tag{3.4}$$

and the potential 2-form [14], [15]

$$\Omega^p = F_{\sigma\beta} dx_\sigma \wedge dx_\beta. \tag{3.5}$$

Postulate 3.1. *The potential 2-form (3.5) is closed, that is*

$$D(\Omega^p) = 0. \tag{3.6}$$

The proof is straightforward. It then follows the set of equations

$$F_{(\sigma\beta\delta)} = \partial_\sigma F_{\beta\delta} + \partial_\delta F_{\sigma\beta} + \partial_\beta F_{\delta\sigma} = 0. \tag{3.7}$$

Another set of equations is obtained from (3.7) for $\sigma = \delta$. In this case,

$$\partial_\sigma F_{\beta\sigma} = \partial_\beta \partial_\sigma A_\sigma - \Box A_\beta. \tag{3.8}$$

The field variables $F_{\beta\sigma}$ are directly measurable under the form of field intensities

$$\begin{cases} \mathbf{E} = -\partial_t \mathbf{A} - \nabla V, \\ \mathbf{B} = \nabla \times \mathbf{A} \end{cases} \tag{3.9}$$

at the point x^β, while the potential 4-vector A_β corresponds to an auxiliar field, determined up to a gauge transformation

$$A_\beta(x) \longrightarrow A_\beta + \partial_\beta \Lambda(x). \tag{3.10}$$

The choice of Λ leads to various gauges. One of the most frequently used is the Lorentz gauge

$$\partial_\sigma A_\sigma = 0. \tag{3.11}$$

Since the 'charge' associated with the field described by A_σ is zero, i.e.

$$\Box A_\beta = 0, \tag{3.12}$$

the set of equations (3.8) reduce to

$$\partial_\sigma F_{\beta\sigma} = 0. \tag{3.13}$$

Equations (3.7) and (3.13) are going to be called *the invariantive equations of the linear field.* The invariance of these equations with respect to transformation (3.10) also yields the condition

$$\Box \Lambda(x) = 0. \tag{3.14}$$

3.2.2. *Minkowskian formulation of motion in the field of particles*

Once the inertial 1-form (2.45) and the potential 1-form (3.1) are introduced, one can study the motion of a particle of 'charge' q (i.e. mass or electric charge) in a linear field.

Postulate 3.2. *Motion of a particle of charge q in the field described by A_β is defined by equality of exterior differentials of 1-forms (2.45) and (3.1), that is*

$$D(\omega^i) = D(\omega^p), \tag{3.15}$$

or, still

$$dp_\beta \wedge dx_\beta = q(dA_\beta \wedge dx_\beta). \tag{3.16}$$

Written explicitly, formula (3.16) becomes

$$(dp_\beta - qF_{\beta\sigma}dx_\sigma) \wedge dx_\beta = 0. \tag{3.17}$$

Multiplying by $(d\tau)^{-1}$, we arrive at the equation of motion

$$\frac{dp_\beta}{d\tau} = qF_{\beta\sigma}u_\sigma; \quad (\beta, \sigma = \overline{1,4}). \tag{3.18}$$

3.2.3. *Electromagnetic and gravitational fields*

Let us identify A_β with the 4-vector potential of the electromagnetic field A_β^e, and q with the electric charge e. Equations (3.7) and (3.13) then yield Maxwell's equations

$$\begin{cases} \nabla \times \mathbf{E}^e = -\partial_t \mathbf{B}^e; & \nabla \cdot \mathbf{B}^e = 0, \\ \nabla \times \mathbf{B}^e = c^2 \partial_t \mathbf{E}^e; & \nabla \cdot \mathbf{E}^e = 0, \end{cases} \tag{3.19}$$

while (3.18) gives rise to the Lorentz force

$$\frac{d\mathbf{p}}{dt} = e(\mathbf{E}^e + \mathbf{v} \times \mathbf{B}^e), \tag{3.20}$$

and the power balance as well

$$\frac{d}{dt}(mc^2) = e\mathbf{E}^e \cdot \mathbf{v}. \tag{3.21}$$

The gravitational field does not accept, in general, a 4-vector potential. Nevertheless, if the field is weak, there exists a solution [16] of the form

$$g_{\alpha\beta} = \eta_{\alpha\beta} + h_{\alpha\beta} \tag{3.22}$$

where $g_{\alpha\beta}$ is the metric tensor, $\eta_{\alpha\beta} = diag[+1, -1, -1, -1]$ is Minkowski's tensor, and $h_{\alpha\beta}$ a small perturbation of the Minkowski metric. With this choice, Einstein's equations

$$R_{\alpha\beta} - \frac{1}{2}Rg_{\alpha\beta} = \frac{8\pi G}{c^4}T_{\alpha\beta}, \tag{3.23}$$

where $R_{\alpha\beta}$ is the Ricci tensor, R the scalar curvature, $T_{\alpha\beta}$ the energy-momentum tensor, G the Newton constant, and c the velocity of light in vacuum for the harmonic gauge conditions

$$\overline{h}^{\alpha\beta}_{,\beta} = \left(h^{\alpha\beta} - \frac{1}{2}h\eta^{\alpha\beta} \right)_{,\beta} = 0;$$

$$\overline{h}_{\alpha\beta} = h_{\alpha\beta} - \frac{1}{2}h\eta_{\alpha\beta}; \quad h = \eta^{\alpha\beta}h_{\alpha\beta} = -\overline{h},$$

(3.24)

take one of the equivalent forms

$$\frac{1}{2}\Box h_{\alpha\beta} - \frac{1}{4}\eta_{\alpha\beta}\Box h = \frac{8\pi G}{c^4}T_{\alpha\beta}; \tag{3.25a}$$

$$\Box \overline{h}_{\alpha\beta} = \frac{16\pi G}{c^4}T_{\alpha\beta}, \tag{3.25b}$$

$$G^{\alpha\beta\gamma}_{,\gamma} \equiv \frac{1}{4}\left(\overline{h}^{\alpha[\beta,\gamma]} + \eta^{\alpha[\beta}\overline{h}^{\gamma]} \right)_{,\gamma} = -\frac{4\pi G}{c^4}T_{\alpha\beta}. \tag{3.25c}$$

Equation (3.25c) stands for a basic equation of analogy between gravitation and electromagnetism.

If the gauge condition is imposed, one obtains

$$G^{\alpha\beta\gamma} = \frac{1}{2}\overline{h}^{\alpha[\beta,\gamma]} \equiv \frac{1}{4}\left(\overline{h}^{\alpha\beta,\gamma} - \overline{h}^{\alpha\gamma,\beta} \right),$$

with properties

$$G^{\alpha[\beta\gamma]} = G^{\alpha\beta\gamma}, \quad G^{[\alpha,\beta\gamma]} = 0, \quad G^{\delta[\alpha,\beta,\gamma]} = 0, \tag{3.26}$$

meaning that $G^{\alpha\beta\gamma}$ represents a gravitational geometric quantity analogous to the electromagnetic field tensor $F^{\alpha\beta}$. Here [...] stands for the anti-symmetrization symbol, e.g.

$$G^{\alpha[\beta,\gamma]} = \frac{1}{2}(G^{\alpha\beta\gamma} - G^{\alpha\gamma\beta}).$$

Let us now introduce the following quantities ([16], [17]):

i) Gravitoelectric scalar potential

$$V_g = -\frac{1}{4}c^2\overline{h}_{00};$$

ii) Gravitomagnetic vector potential

$$\mathbf{A}_g = (A_g^i) = \frac{1}{4}c^2\overline{h}_{0i}, \quad (i = \overline{1,3});$$

iii) Gravitoelectric field intensity

$$\mathbf{E}_g = (E_g^i) \equiv (c^2 G^{00i}) = -c^{-1}\partial_t\mathbf{A}_g - \nabla V_g;$$

iv) Gravitomagnetic field intensity

$$\mathbf{B}_g = \nabla \times \mathbf{A}_g;$$

v) Energy-momentum tensor

$$T^{\alpha\beta} = \rho_g u^\alpha u^\beta = \rho_{mass} u^\alpha u^\beta; \quad u^\alpha u_\alpha = c^2;$$

$$U^\alpha = \frac{dx^\alpha}{d\tau} = \left(1 - \frac{v^2}{c^2}\right)^{-1/2}\frac{dx^0}{dt};$$

$$x^0 = ct, \quad x^1 = x, \quad x^2 = y, \quad x^3 = z.$$

vi) Small velocities approximation for the energy-momentum tensor. In this case, the mixed spatial components of the energy-momentum tensor (tension components), and the terms quadratic in v^2 are neglected (even if the translation and rotation velocities remain high enough). Within this approximation, the only non-zero components of the energy-momentum tensor are

$$T^{00} = \rho_g c^2; \quad T^{0i} = \rho_g c\frac{dx^i}{dt} = \rho_g cv^i.$$

Introducing these quantities into (3.25c) and (3.26), the gravito-electromagnetism equations are obtained:

$$\begin{cases} \nabla\mathbf{E}_g = -\partial_t\mathbf{B}_g; \quad \nabla \cdot \mathbf{E}_g = -4\pi G\rho_g, \\[2ex] \nabla \times \mathbf{B}_g = -\dfrac{4\pi G}{c}\rho_g\mathbf{v} + \dfrac{1}{c}\partial_t\mathbf{E}_g; \quad \nabla \cdot \mathbf{B}_g = 0. \end{cases} \tag{3.27}$$

The harmonic gauge condition (within the linear approximation) becomes a Lorentz-type condition for gravito-electromagnetic potentials

$$\nabla \cdot \mathbf{A}_g + \frac{1}{c}\partial_t V_g = 0.$$

The gauge condition (3.24) could be used to obtain a vector gauge condition

$$\partial_t \mathbf{A}_g = 0,$$

but, if used, it could lead to the inexistence of both gravitomagnetic induction

$$\partial_t \mathbf{B}_g = 0$$

and gravitomagnetic dynamo phenomenon (amplification of a seed intrinsic rotation or spin motion by gravitomagnetic induction). Therefore, due to the fact that a "magnetic" Galilean limit for the gravitomagnetic equations is physically inconsistent, there is no justification for the use of such a gauge in the general case.

The analogy between gravitation and electromagnetism is also put into evidence by the following relations, obtained for a stationary (the gravitational potentials are time-independent), linear gravitational field:

i) $$h^{00}_{,0} = \left(h^{00} - \tfrac{1}{2}\eta^{00}h\right)_{,0} = 0; \quad \overline{h}^{0i}_{,0} = 0,$$
$$\partial_t V_g = 0; \qquad \partial_t \mathbf{A}_g = 0.$$

ii) Gravito-electromagnetostatics equations

$$\nabla \times \mathbf{E}_g = 0; \qquad \nabla \cdot \mathbf{E}_g = -4\pi G\rho_g,$$
$$\nabla \times \mathbf{B}_g = -\frac{4\pi G}{c}\rho_g\mathbf{v}; \qquad \nabla \cdot \mathbf{B}_g = 0.$$

iii) Coulomb-type gauge condition

$$\nabla \cdot \mathbf{B}_g = 0.$$

iv) The gravito-electric field intensity is identical to the Newtonian acceleration

$$\mathbf{E}_g \equiv \mathbf{g} = -\nabla V_g.$$

v) The gravito-magnetic Lorentz force is obtained by means of the geodesic equations

$$\frac{d^2 x^\alpha}{d\tau^2} + \Gamma^\alpha_{\beta\gamma} \frac{dx^\beta}{d\tau} \frac{dx^\gamma}{d\tau} = 0,$$

or

$$\frac{d\mathbf{u}}{d\tau} = \mathbf{E}_g + \frac{4}{c}(\mathbf{u} \times \mathbf{B}_g),$$

where $\mathbf{u}$ is the velocity of the test particle introduced into the gravito-electromagnetic field $(\mathbf{E}_g, \mathbf{B}_g)$. The factor 4 comes from a scale difference between the gravito-electromagnetic force and the electromagnetic Lorentz force.

vi) The gravito-magnetic source field $\mathbf{S}$

$$\mathbf{A}_g = -\frac{1}{2} \frac{\mathbf{S} \times \mathbf{r}}{cr^3},$$

$$\mathbf{B}_g = \nabla \times \mathbf{A}_g = -\frac{1}{2} \frac{3\mathbf{n}(\mathbf{S} \cdot \mathbf{n}) - \mathbf{n}}{cr^3}; \quad \mathbf{n} = \frac{\mathbf{r}}{r}.$$

As it can be observed, these equations are analogous to those corresponding to the field produced by the magnetic dipole. The only difference is that the magnetic dipole moment is replaced by angular momentum times $(-1/2)$.

In the same context, the gravitomagnetic field induced by the Earth rotation, at 45° geographic latitude, is given by

$$|\mathbf{B}_g| = \left| \frac{GI}{2cR^3} \left[\boldsymbol{\omega} - \frac{3(\boldsymbol{\omega} \cdot \mathbf{R})\mathbf{R}}{R^2} \right] \right| \approx 20.4 \times 10^{-14} s^{-1},$$

with R — the Earth radius, I — the inertia moment of the Earth rotating about an axis passing through its center, and ω — the angular velocity of rotation of the Earth.

The linear approximation is appropriate at the surface of any component of our Solar System. Here $\mathbf{E}_g$ represents the Newtonian gravitational acceleration, containing first-order corrections to flat space, while $\mathbf{B}_g$ is connected to interaction of angular momenta and contains second-order corrections. Effects due to $\mathbf{B}_g$ are 10^{12} times smaller than those attributed to $\mathbf{E}_g$.

Even on surface of a white dwarf star (a star with radius $R \geq 1 \cdot 10^6 m$, and mass comparable with the mass of the Sun, $M_\odot = 2 \cdot 10^{30} kg$, which is able to rotate about its own axis, but not as rapidly as a neutron star — with radius one hundred times smaller — that can represent a pulsar component), deviation of g_{00} from unity (i.e. with respect to the flat space metric) is about 10^{-4}.

Consequently, one can postulate *ab initio* Maxwell's type gravitational equation to describe the gravitational field. In this case, according to our previous observations, one can admit a 4-vector potential A_β^g and the theory developed in paragraph 3.2.1 can be applied.

3.2.4. *Anti-Maxwellian-type fields*

Since while studying inertial motion of the material point the value $\epsilon = -1$ has been admitted, one can also consider (within the frame of linear invariantive field theories) a field whose potentials $(\mathbf{A}_\omega, V_\omega)$, called *anti-Maxwellian* by Octav Onicescu ([4].[5]), satisfy the equations

$$\nabla \cdot \mathbf{A}_\omega + \frac{1}{\omega}\frac{\partial V_\omega}{\partial t} = 0;$$

$$\Delta \mathbf{A}_\omega + \frac{1}{\omega^2}\frac{\partial^2 \mathbf{A}_\omega}{\partial t^2} = 0, \tag{3.28}$$

and define the field intensities

$$\mathbf{E}_\omega = -\nabla V_\omega - \frac{\partial \mathbf{A}_\omega}{\partial t}; \qquad \mathbf{B}_\omega = \nabla \times \mathbf{A}_\omega. \tag{3.29}$$

The fields $(\mathbf{E}_\omega, \mathbf{B}_\omega)$ satisfy the equations

$$\nabla \times \mathbf{E}_\omega + \frac{\partial \mathbf{B}_\omega}{\partial t} = 0; \tag{3.30}$$

$$\nabla \cdot \mathbf{B}_\omega = 0; \tag{3.31}$$

$$\nabla \times \mathbf{B}_\omega + \frac{1}{\omega^2}\frac{\partial \mathbf{E}_\omega}{\partial t} = 0; \tag{3.32}$$

$$\nabla \cdot \mathbf{E}_\omega = 0. \tag{3.33}$$

As one can see, equations (3.30), (3.31), and (3.33) are similar to Maxwell's equations, but (3.32) is different.

The essential difference between Maxwellian and anti-Maxwellian fields becomes obvious by comparing (3.12) and (3.28) for $\omega = c$. Equation (3.12) describes a phenomenon of wave propagation, while (3.28) imply a "global internal tension": the waves cannot "detach" from the body, and this fact giver rise to a field of tensions.

From the perspective of transformations induced by the bilinear form ϕ and the Euclidean metric $d\sigma^2$ (see § 2.3), equations (3.30) — (3.33) describe the *instantonic* field. Such a field spontaneously breaks the symmetry of the electrono-positronic vacuum by the mechanisms of tunneling, generating 'charge' (mass, or electric charge).

3.3. Nonlinear field theory

Consider the inertial 1-form

$$\omega^i = p_\beta dx^\beta \tag{3.34}$$

together with the kinematic one

$$\omega^c = u_\beta dx^\beta, \tag{3.35}$$

where p_β and u_β are the momentum and velocity 4-vectors, respectively, with

$$u_\beta = \frac{dx_\beta}{d\tau}; \quad (\beta = \overline{1,4}).$$

Postulate 3.3. *For a free material point, the exterior product of the 1-forms (3.34) and (3.35) is null, that is*

$$\omega^i \wedge \omega^c = 0. \tag{3.36}$$

It then implicitly follows that

$$\mathbf{p} = m\mathbf{v}; \tag{3.37}$$

$$E = mc^2. \tag{3.38}$$

But in arbitrary conditions a particle does not obey relation (3.36), so that the field 2-form

$$\Omega^f = A_{\alpha\beta}dx^\alpha \wedge dx^\beta \tag{3.39}$$

has to be introduced. In general, this 2-form is not an exterior product.

Postulate 3.4. *The arbitrary motion of a particle is given by equality between the exterior differential of the inertial 1-form (3.34) and the field 2-form (3.39), that is*

$$D(\omega^i) = \Omega^f. \tag{3.40}$$

Explicitly, this can be written as

$$(dp_\sigma - qA_{\sigma\beta}dx^\beta) \wedge dx^\sigma = 0, \tag{3.41}$$

or, after multiplying by $(d\tau^{-1})$

$$\frac{dp_\sigma}{d\tau} = qA_{\sigma\beta}u^\beta. \tag{3.42}$$

The 16-component tensor field $A_{\sigma\beta}$ generalizes the force of classical mechanics.

The necessity of defining such a tensor was put into evidence by Octav Onicescu ([4], [5]). Observing that the mechanical work given by

$$dL = X_i dx_i \qquad (i = \overline{1,3})$$

is no longer an exact total differential and using the principles of invariantive mechanics, he introduces the 4-potential variation

$$dA_\beta = A_{\beta\sigma}dx^\sigma, \qquad (\beta, \sigma = \overline{1,4})$$

where $A_{\beta\sigma}$ is a 16-component tensor. The law of motion is given by

$$D(\Omega^f) = D(A_{\beta\sigma} dx^\beta \wedge dx^\sigma) = 0.$$

Postulate 3.5. *The field 2-from Ω^f is closed, that is*

$$D(\Omega^f) = 0. \tag{3.43}$$

Explicitly, this relation yields

$$A_{(\mu\sigma\beta)} = \partial_\beta A_{\mu\sigma} + \partial_\sigma A_{\beta\mu} + \partial_\mu A_{\sigma\beta} = 0. \tag{3.44}$$

If $A_{\sigma\beta} = F_{\sigma\beta}$, which implies existence of the potential 4-vector A_σ, then (3.44) reduce to the set of equations (3.7).

According to one of Cartan's theorems [14], if (3.41) is true, then

$$(dp_\sigma - A_{\sigma\beta} dx^\beta) \wedge dx^\sigma = -q_{\sigma\beta} dx^\beta \wedge dx^\sigma, \tag{3.45}$$

or, still,

$$dp_\sigma = (A_{\sigma\beta} - g_{\sigma\beta}) dx^\beta, \tag{3.46}$$

where $g_{\sigma\beta}$ is a symmetric tensor. If (3.46) is multiplied by $(d\tau)^{-1}$, we are left with the equation of motion

$$\frac{dp_\sigma}{d\tau} = (A_{\sigma\beta} - g_{\sigma\beta}) u^\beta. \tag{3.47}$$

The difference $(A_{\sigma\beta} - g_{\sigma\beta})$ has to be a tensor.

Let us now define the *Hilbert's tensor* $T^{\sigma\beta}$ associated with $A_{\sigma\beta}$, also called *constraint tensor*, by

$$T^{\sigma\beta} = \sqrt{I}\, g^{\sigma\beta}, \tag{3.48}$$

with

$$I = \det(\delta_\beta^\sigma - A_\beta^\sigma). \tag{3.49}$$

If $g_{\sigma\beta}$ is a metric tensor, then $T^{\sigma\beta}$ can be identified with the energy-momentum tensor only if

$$\nabla_\beta T^{\sigma\beta} = 0, \tag{3.50}$$

where ∇_β is the covariant derivative within the natural connection induced by $g_{\sigma\beta}$. Explicitly, (3.50) writes

$$\nabla_\beta T_\mu^\beta = A_\mu^\sigma \nabla_\beta(\sqrt{I} A_\sigma^\beta) + \frac{1}{2}\sqrt{I} A^{\beta\sigma} A_{(\mu\beta\sigma)}. \tag{3.51}$$

Here, the following notation has been used:

$$A_{(\mu\beta\sigma)} = \nabla_\mu A_{\beta\sigma} + \nabla_\sigma A_{\mu\beta} + \nabla_\beta A_{\sigma\mu}. \tag{3.52}$$

Since, according to (3.44), $A_{(\mu\beta\sigma)} = 0$, we still have

$$\nabla_\beta T_\mu^\beta = A_\mu^\sigma \nabla_\beta(\sqrt{I} A_\sigma^\beta), \tag{3.53}$$

or, in view of (3.50)

$$\nabla_\beta(\sqrt{I} A_\sigma^\beta) = 0, \tag{3.54}$$

where the condition

$$\det(A_\beta^\sigma) \neq 0 \tag{3.55}$$

has been imposed.

Equations (3.54) are identical with the Born-Infeld equations of non-linear electrodynamics and, in the context of this discussion, they express the law of conservation of the energy-momentum tensor (3.48).

We shall call equations (3.44) and (3.50) *equations of invariantive mechanics for the non-linear field*. These equations allow us to both determine the tensor $A_{\sigma\beta}$ and give a correct definition of the energy-momentum tensor $T^{\sigma\beta}$.

3.4. Comments of Maxwellian and anti-Maxwellian type fields

The above analysis shows that the anti-Maxwellian type fields (i.e. electromagnetic and gravitational within linear approximation) express the stability of material structures at the microscopic scale (stability of elementary particles, etc.), as well as at the macroscopic scale (stability of cosmic structures like binary stars, binary galaxies, etc.). This is explained by the fact that such fields do not propagate, but remain localized on structures as internal tensions. This way, both individuality and stability of structures are assured.

On the other hand, the Maxwellian-type fields imply, from the point of view of propagation of electromagnetic and/or gravitational waves, the presence of both transversal and longitudinal components (even if, usually, we are tempted to consider only one component, usually the transversal one — see [17] for details). The presence of various field components could be explained supposing, for example, that very closed to a black hole the gravitational field is very high, while at large distances from the black hole the gravitational and electric (corresponding to an electrically charged black hole) fields are analogous to those produced by a usual (charged) star. There arises a question: since no kind of particles (gravitons, photons, etc.) can escape from a black hole, how could be possible for a black hole to interact (gravitationally or electromagnetically) with some other astrophysical objects?

The answer is offered by quantum field theory (QFT) [4]. As well known, in electromagnetism and gravito-electromagnetism (linear gravitation) the field physical quantities have two components: i) The transversal components, in which case momentum and energy form a 4-vector which behaves like momentum and energy of the particles; ii) longitudinal (or scalar) component. In QFT only the transversal component is quantized, the corresponding quanta being the gravitons (for gravitational field) and photons (for electromagnetic field). The scalar non-quantized part refers to the electrostatic (Coulombian) or gravitostatic (Newtonian) potential.

The longitudinal gravitons are able to leave the black hole, but the transversal gravitons, which correspond in fact to the gravitational waves, cannot do it. The situation is similar to that encountered in the electromagnetic field. The electromagnetic radiation involves transversal (observable) photons, which are energy carriers, while the Coulombian field contains longitudinally polarized (virtual) photons that do not carry energy and cannot be observed as free particles. Consequently, there is a direct interaction between a transversal photon and the gravitational field of a black hole, but there is no interaction between a longitudinal photon and a black hole. In other words, the Coulombian field (the electrostatic interaction by means of force lines) can intersect the event horizons of a black hole (or, by approximation, its surface).

A similar distinction can be considered between longitudinal and transversal gravitons. A black hole can attract both matter and radiation outside its event horizon, since the gravitational field is 'carried' by the longitudinal gravitons. Sometimes it is stated that, in statical case, the electric and gravitational field are quantized by photons, respectively gravitons, which are time-polarized, or along their direction of motion. Therefore, the photons and gravitons with such polarizations are able to transmit the static fields beyond the event horizon. Now, another legitimate question can arise: are the photons (gravitons) time-polarized able to indicate a time arrow?

Such questions and concepts can be developed in a more general frame, in which the Universe itself or an elementary particle can be considered as black holes. The Birkhoff, Gauss and "no-hair" theorems show that the static gravitational and electric fields are able to leave the black hole.

Chapter 4

Ondulatory Invariantive Theories. Wave-Corpuscule Duality

4.1. Preliminaries

In this chapter we shall deal with the linear wave theory, the linear de Broglie theory, and de Broglie-Takabayasi theory as well [6,13]. As the starting point, we shall take advantage of the studies concerning the photon and the wave associated to a material point, first developed in invariantive mechanics by Octav Onicescu [5].

As in the previous chapters, we shall characterize the wavelike motion by a 1-form whose closeness leads to the canonical equations in ondulatory formalism. Louis de Broglie's principle, according to which a wave can be associated to any particle in motion, shall be expressed by means of equivalence between an inertial and an ondulatory 1-forms. Using the wave equation for scalar particles with non-zero rest mass, the de Broglie-Takabayas equation is re-obtained.

4.2. Linear wave theory

Let us consider a wave characterized by angular frequency ω and wave vector $\mathbf{k}$, and assume to describe this wave by the ondulatory 1-form

$$\omega^o = \mathbf{k}d\mathbf{r} - \omega dt. \tag{4.1}$$

Postulate 4.1. *Motion of the free wave is given by the closeness of the 1-form (4.1), that is*

$$D(\omega^o) = d\mathbf{k} \wedge d\mathbf{r} - d\omega \wedge dt = 0, \tag{4.2}$$

or, explicitly,

$$D(\omega^o) = d\mathbf{k}\delta\mathbf{r} - \delta\mathbf{k}d\mathbf{r} - d\omega\delta t + \delta\omega dt = 0. \tag{4.2'}$$

Since $\omega = \omega(\mathbf{k}, \mathbf{r})$, we can write

$$\delta\omega = \frac{\partial\omega}{\partial\mathbf{k}}\delta\mathbf{k} + \frac{\partial\omega}{\partial\mathbf{r}}\delta\mathbf{r}. \tag{4.3}$$

A convenient arrangements of terms in (4.2) yields

$$D(\omega^o) = \left(d\mathbf{k} + \frac{\partial\omega}{\partial\mathbf{r}}dt\right)\delta\mathbf{r} + \left(-d\mathbf{r} + \frac{\partial\omega}{\partial\mathbf{k}}dt\right)\delta\mathbf{k} - d\omega\delta t = 0. \tag{4.4}$$

Whereas variations $\delta\mathbf{r}$, $\delta\mathbf{k}$, δt are arbitrary, relation (4.4) implies the following equations in ondulatory formalism:

i) Canonical equations

$$\dot{\mathbf{k}} = -\frac{\partial\omega}{\partial\mathbf{r}}; \quad \dot{\mathbf{r}} = \frac{\partial\omega}{\partial\mathbf{k}} \tag{4.5}$$

where $\dot{\mathbf{r}}$ stands for the group velocity.

ii) Equation of conservation of energy

$$\frac{d\omega}{dt} = 0. \tag{4.6}$$

If the medium is isotropic, one can assume

$$\omega = \omega\left(\frac{1}{2}\mathbf{k}^2\right), \tag{4.7}$$

in which case (4.5) become

$$\dot{\mathbf{k}} = 0, \tag{4.8}$$

and

$$\mathbf{v} = \frac{\partial \omega}{\partial \mathbf{k}^2}\mathbf{k}. \tag{4.9}$$

Choosing

$$\frac{\partial \omega}{\partial \mathbf{k}^2} = \mu = const. \tag{4.10}$$

we finally have

$$\mathbf{v} = \mu\mathbf{k} \tag{4.11}$$

which would correspond to longitudinal waves [18].

Relation

$$v = \frac{\omega}{k}, \tag{4.12}$$

would define the phase velocity of the wave.

Therefore, to a wave one can associate two types of velocities:

i) Group velocity, which identifies with the velocity of propagation of the wave;
ii) Phase felocity, with no straight physical signification.

Within the wavelike formalism, the Lagrangian function is

$$L = \mathbf{k}\frac{\partial \omega}{\partial \mathbf{k}} - \omega, \tag{4.13}$$

If the group and the phase velocities are equal, then $L = 0$.

4.3. Minkowskian formulation of the ondulatory linear theory

Using the ondulatory 1-form (4.1), weitten as

$$\omega^o = \mathbf{k}d\mathbf{r} + \frac{i}{c}\omega icdt = k_\beta dx_\beta \quad (\beta = \overline{1,4}), \tag{4.14}$$

one can define the wave 4-vector

$$k_\beta = \left(\mathbf{k}, \frac{i}{c}\omega\right). \tag{4.15}$$

Postulate 4.2. *The wave 4-dimensional motion is given by closeness of the 1-form (4.14), that is*

$$D(\omega^o) = dk_\beta \wedge dx_\beta = 0, \tag{4.16}$$

or, explicitely,

$$dk_\beta \delta x_\beta - \delta k_\beta dx_\beta = 0. \tag{4.17}$$

Multiplying (4.17) by $(d\tau)^{-1}$ and equating to zero the coefficient of δx_β, we are left with the law of conservation of the wave 4-vector

$$\frac{dk_\beta}{d\tau} = 0, \tag{4.18}$$

as well as the structure law

$$u_\beta \delta k_\beta = 0. \tag{4.19}$$

Relation (4.18) expresses the ondulatory formulation of the 4-momentum conservation law, while (4.19) displays the orthogonalyty of the 4-vectors u_β and δk_β,

Since the Lagrangian (4.13) is zero, that is

$$k_\beta k_\beta = 0, \tag{4.20}$$

the choice

$$k_\beta = \partial_\beta \psi' \tag{4.21}$$

allows us to write the eikonal equation

$$\partial_\beta \psi' \partial_\beta \psi' = 0, \tag{4.22}$$

where ψ' is the eikonal function.

4.4. De Broglie's linear theory

Consider the inertial 1-form

$$\omega^i = p_\beta dx_\beta, \quad (\beta = \overline{1,4}) \tag{4.23}$$

as well as the ondulatory 1-form

$$\omega^o = k_\beta dx_\beta. \tag{4.24}$$

Postulate 4.3. *Admitting that to any moving particle one can associate a wave (de Broglie's hypothesis on wave-cospuscule duality), then the exterior product of the 1-forms (4.23) and (4.24) is null, that is*

$$\omega^i \wedge \omega^o = 0. \tag{4.25}$$

According to (4.25) the 1-forms ω^i and ω^o are proportional, *i.e.*

$$p_\beta = \mu' k_\beta, \quad \mu' = const. \tag{4.26}$$

If

$$\mu' = \frac{h}{2\pi} = \hbar,$$

where h is Planck's constant, relations (4.26) take the form of de Broglie's relations

$$\mathbf{p} = \hbar\mathbf{k}; \quad E = h\nu. \tag{4.27}$$

Taking into account relation between Planck's constant and Onicescu's informational energy (see Chap.11 for details), it then follows from (4.25) that the duality wave-corpuscule fulfils for the same value of the informational energy.

The Hamilton-Jacobi equation emerges from

$$p_\beta p_\beta = -m_0^2 c^2 \qquad (4.28)$$

with

$$p_\beta = \partial_\beta S \qquad (4.29)$$

where S is the action. It follows

$$\partial_\beta S \partial_\beta S + (m_0 c)^2 = 0, \qquad (4.30)$$

or, still

$$\partial_i S \, \partial_i S - c^{-2}(\partial_t S)^2 + (m_0 c)^2 = 0; \quad (i = \overline{1,3}). \qquad (4.31)$$

Let us homogenize equation (4.31) by means of some function

$$\chi = \chi(S, x_i, t), \qquad (4.32)$$

which has to be null when S verifies (4.31). In this case

$$\partial_i S = (\partial_\chi S)(\partial_i \chi) \quad \partial_t S = (\partial_\chi S)(\partial_t \chi),$$

and (4.31) becomes

$$\partial_i \chi \partial_i \chi - c^{-2}(\partial_t \chi)^2 + (m_0 c)^2 (\partial_S \chi)^2 = 0. \qquad (4.33)$$

In the stationary case, we have

$$\partial_t \chi = E \partial_S \chi, \qquad (4.34)$$

and (4.33) writes

$$\partial_i \chi \partial_i \chi = \frac{v^2}{c^4}(\partial_t \chi)^2 = 0. \qquad (4.35)$$

The corresponding wave equation is

$$\Delta \Phi - \frac{v^2}{c^4} \partial_{tt} \Phi = 0. \tag{4.36}$$

This way is put into evidence the wave propagation velocity

$$w = \frac{c^2}{v}, \tag{4.37}$$

that is de Broglie's relation between the two species of velocity: the wave velocity w, and the corpuscle velocity v. Using this observation, one can further obtain Schrödinger's equation and, therefore, the quantization. This way, a differential operator with specific properties one associates to each physical quantity.

4.5. The de Broglie-Takabayashi theory

Let us first construct a differential equation for a particle possessing the rest mass m_0 and characterized by the wave function Ψ. To this end, in equation (4.28) we replace the 4-momentum p_β by the corresponding operator

$$p_\beta \to i\hbar \partial_\beta, \tag{4.38}$$

leading to Klein-Gordon equation

$$\Box \Psi = \left(\frac{m_0 c}{\hbar}\right)^2 \Psi, \tag{4.39}$$

which is fundamental for our further investigations.

4.5.1. *The de Broglie-Takabayashi equations*

Let us choose in (4.30) function Ψ of the form

$$\Psi = A e^{\frac{i}{\hbar} S}, \tag{4.40}$$

where A is the amplitude, and S the phase. Separating the imaginary and real parts, we are left with

$$\partial_\beta[A^2 \partial_\beta(S)] = 0; \tag{4.41}$$

$$\partial_\beta S \, \partial_\beta S + m_0^2 c^2 - \hbar^2 \frac{\Box A}{A} = 0. \tag{4.42}$$

In this context, relation

$$u_\beta = m_0^{-1} \partial_\beta S \tag{4.43}$$

cannot be used to define the velocity 4-vector, because the condition

$$u_\beta u_\beta = -c^2 \tag{4.44}$$

is not generally verified. To solve the problem, one introduces the fluctuant proper mass M_0 [19]

$$M_0 = \left(m_0^2 - \frac{\hbar^2}{c^2} \frac{\Box A}{A} \right)^{1/2}, \tag{4.45}$$

and (4.42) becomes

$$\partial_\beta S \, \partial_\beta S + M_0^2 c^2 = 0.$$

In this case, equation (4.44) is satisfied by taking

$$u_\beta = M_0^{-1} \partial_\beta S \tag{4.46}$$

while (4.42) becomes

$$M_0 u_\beta \partial_\beta S = M_0 \dot{S} = -M_0^2 c^2 = -E_0 M_0, \tag{4.47}$$

where

$$E_0 = M_0 c^2 \tag{4.48}$$

is called *fluctuant proper energy*.

Next, if we define the fluctuant density

$$\rho = A^2 \frac{M_0}{m_0},\tag{4.49}$$

then (4.41) shall lead to the law of conservation of the current density j_β

$$\partial_\beta j_\beta = 0,\tag{4.50}$$

with

$$j_\beta = \rho u_\beta,\tag{4.51}$$

while (4.42) gives the equation of motion

$$\frac{d}{d\tau}(M_0 u_\beta) = -\frac{\partial}{\partial x_\beta}(M_0 c^2).\tag{4.52}$$

Multiplying by ρ and using the identities

$$\rho \partial_\tau (M_0 u_\beta) = \partial_\tau (\rho M_0 u_\beta) - M_0 u_\beta \partial_\tau \rho$$
$$= \partial_\alpha (\rho M_0 u_\alpha u_\beta) - M_0 u_\beta \partial_\alpha (\rho u_\alpha) = \partial_\alpha (\rho M_0 u_\alpha u_\beta);\tag{4.53}$$
$$-\rho \partial_\beta (M_0 c^2) = -A^2 \frac{M_0}{m_0} \partial_\beta (M_0 c^2) = \frac{\hbar^2 A^2}{2m_0} \partial_\beta \left(\frac{\Box A}{A}\right)$$
$$= \frac{\hbar^2}{2m_0}[A\partial_\beta(\Box A) - \partial_\beta A(\Box A)] = \frac{\hbar^2}{2m_0}\partial_\alpha[(A\partial_\beta \partial_\alpha A)$$
$$- \partial_\beta A \partial_\alpha A] = \partial_\alpha \left[\frac{\hbar^2 A^2}{2m_0}\partial_\beta \partial_\alpha (\log A)\right],\tag{4.54}$$

one obtains the equation of conservation of energy-momentum tensor

$$\partial_\alpha t_{\alpha\beta} = 0,\tag{4.55}$$

where

$$t_{\alpha\beta} = \rho M_0 u_\alpha u_\beta - \rho \frac{\hbar^2}{2M_0}\partial_\alpha \partial_\beta (\log A).\tag{4.56}$$

It follows that:

i) Any particle interacts with a 'sub-quantum medium' of density ρ (4.49);

ii) Interaction between the particle and the 'sub-quantum medium' is dictated by the quantum potential

$$Q = M_0 c^2; \tag{4.57}$$

iii) As a result of this interaction, the particle gets a fluctuant proper mass M_0 (4.45), and a fluctuant proper energy E_0 (4.48);

iv) The 'sub-quantum medium' identifies with a quantum fluid, described by the equations de Broglie-Takabayashi (4.50) and (4.56);

v) By means of the 'instanton' [20]

$$I = -\ln \Psi, \tag{4.58}$$

one introduces the generalized 4-velocity

$$V_\beta = -i\hbar M_0^{-1} \partial_\beta I = u_\beta + i v_\beta, \tag{4.59}$$

with

$$u_\beta = M_0^{-1} \partial_\beta S; \quad v_\beta = \hbar M_0^{-1} \partial_\beta \ln A. \tag{4.60}$$

Equations (4.41) and (4.55) then become

$$u_\beta v_\beta = -\frac{\hbar^2}{M_0^2} \Box S, \tag{4.61}$$

respectively

$$\partial_\alpha \left[\rho M_0 u_\alpha u_\beta - \frac{\rho \hbar}{2} \partial_\alpha v_\beta \right] = 0; \tag{4.62}$$

vi) For $M_0 = m_0$, condition

$$\partial_\beta V_\beta = 0 \qquad (4.63)$$

implies

$$\partial_\beta u_\beta = \square S = 0; \quad \partial_\beta v_\beta = \square(\ln A) = 0, \qquad (4.64)$$

meaning that both the phase and logarithm of the amplitude propagate with velocity c. In this case the de Broglie-Takabayashi quantum fluid becomes incompressible ($\partial_\tau \rho = -\square S = 0$), and the 4-velocity fields u_β and v_β become orthogonal ($u_\beta v_\beta = 0$).

4.5.2. *Non-relativistic approximation of the de Broglie-Takabayashi theory. The hydrodynamic model of quantum mechanics*

The correspondence between relativistic action S and classical action S' is given by

$$S' = S + m_0 c^2 t. \qquad (4.65)$$

Then, by means of identities

$$\partial_t S = \partial_t S' - m_0 c^2, \quad \partial_{tt} S = \partial_{tt} S';$$

$$\partial_\beta S \, \partial_\beta S = (\nabla S')^2 - c^{-4}(\partial_t S' - m_0 c^2)^2$$

$$\approx (\nabla S')^2 + 2 m_0 \partial_t S' - m_0^2 c^2;$$

$$\square S = \Delta S'; \quad \square A = \Delta A; \qquad (4.66)$$

$$\partial_\beta(A^2 \partial_\beta S) = \nabla A^2 \nabla S' - c^{-2}(\partial_t S' - m_0 c^2)\partial_t A^2 + A^2 \Delta S'$$

$$\approx \nabla A^2 \nabla S' + m_0 \partial_t(A^2) + A^2 \Delta S',$$

equations (4.41) and (4.42) write

$$\partial_t(m_0 A^2) + \nabla(A^2 \nabla S') = 0;$$

$$2 m_0 \partial_t S' + (\nabla S')^2 - \hbar^2 \frac{\Delta A}{A} = 0, \qquad (4.67)$$

If, alternatively, one defines

$$\mathbf{v} = m_0^{-1}\nabla S'; \quad \rho = A^2, \tag{4.68}$$

then (4.41) and (4.42) yield

$$\partial_t \rho + \nabla \cdot (\rho \mathbf{v}) = 0;$$
$$m_0[\partial_t \mathbf{v} + (\mathbf{v} \cdot \nabla)\mathbf{v}] = -\nabla Q, \tag{4.69}$$

where

$$Q = -\frac{\hbar^2}{2m_0}\frac{\Delta \rho^{1/2}}{\rho^{1/2}}. \tag{4.70}$$

This way, we have encountered the hydrodynamic model of quantum mechanics [19].

An equivalent formulation is obtained by introducing the complex velocity

$$\mathbf{V} = -\frac{i\hbar}{m_0}\nabla(\ln \psi) = \mathbf{v} + i\mathbf{u}, \tag{4.71}$$

where

$$\psi = Ae^{\frac{i}{\hbar}(S' - m_0 c^2 t)};$$
$$\mathbf{v} = \frac{\nabla S'}{m_o};$$
$$\mathbf{u} = -\frac{\hbar}{2m_0}\nabla(\ln \rho). \tag{4.72}$$

The resulting equations are equivalent with (4.69)

$$\partial_t \rho + \nabla \cdot (\rho \mathbf{v}) = 0;$$
$$m_0 \partial_t \mathbf{v} = -\nabla\left(\frac{m_0 \mathbf{v}^2}{2} + Q\right), \tag{4.73}$$

where

$$Q = -\frac{\hbar}{2}\nabla \cdot \mathbf{u} - \frac{m_o \mathbf{u}^2}{2}. \tag{4.74}$$

These equations define the quantum hydrodynamic model (for details see Chap. 13).

4.6. Hamilton-Jacobi equation associated with the motion of a particle in various fields. Correspondence with de Broglie's theory

In the presence of a linear electromagnetic or gravitational field, characterized by potentials $\mathbf{A}_l, V_l$ $(l = e, g)$, the linear momentum $\mathbf{p}$ and the energy H of a free particle

$$\mathbf{p} = m\mathbf{v} = \frac{H}{c^2}\mathbf{v},$$

$$H = c(m_0^2 c^2 + \mathbf{p}^2)^{1/2} \tag{4.75}$$

change, the generalized momentum and energy being, respectively

$$\mathbf{p}_l = \mathbf{p} + q_l \mathbf{A}_l;$$

$$H_l = c[m_0^2 c^2 + (\mathbf{p}_l - q_l \mathbf{A}_l)^2]^{1/2} + q_l V_l. \tag{4.76}$$

The Hamilton-Jacobi equation of particle in these fields is obtained by means of substitutions

$$p_l = \nabla S; \quad H_l = \partial_t S, \tag{4.77}$$

which yield

$$(\nabla S - q_l \mathbf{A}_l)^2 - c^2(\partial_t S + q_l V_l)^2 + m_0^2 c^2 = 0. \tag{4.78}$$

This equation can be homogeneized through the medium of function

$$\chi = \chi(S, \mathbf{r}, t) \tag{4.79}$$

which has to be zero when S verifies (4.78). Then

$$\nabla S = \partial_\chi S \nabla \chi; \quad \partial_t S = \partial_\chi S \partial_t \chi$$

shall satisfy equation

$$(\nabla \chi)^2 - c^{-2}(\partial_t \chi)^2 + q_l^2(\mathbf{A}_l^2 - c^{-2}V_l^2 + q_l^{-2} m_0^2 c^2)(\partial_S \chi)^2$$

$$-2q_l(\mathbf{A}_l \nabla \chi - c^{-2}V_l \partial_t \chi)(\partial_S \chi) = 0. \tag{4.80}$$

(Here, as elsewhere, the dot product sign has been omitted.)

If S is considered as being stationary (see 4.34),

$$\partial_t \chi = E \partial_S \chi,$$

equation (4.80) becomes

$$(\nabla \chi)^2 + (q_l^2 \mathbf{A}_l^2 - \mathbf{p}^2)(\partial_S \chi)^2 - 2 q_l \mathbf{A}_l \nabla \chi (\partial_S \chi) = 0, \qquad (4.81)$$

which yields the characteristics of the wave equation

$$\Delta \phi + (q_l^2 \mathbf{A}_l^2 - \mathbf{p}^2)\partial_{SS}\phi - 2 q_l \mathbf{A}_l \nabla (\partial_S \phi) = 0. \qquad (4.82)$$

From now on, by means of the method used in paper [5], a quantization theory can be developed.

We are now able to make connection to de Broglie's theory of the double solution [21, 22], on the basis of relations (4.75), (4.76) and (4.77). If the fields are absent, one obtains

$$\mathbf{v} = \frac{c^2}{H}\mathbf{p} = -c^2 \frac{\nabla S}{\partial_t S}, \qquad (4.83)$$

while the presence of a field gives

$$\mathbf{v} = -c^2 \frac{\nabla S + q_l \mathbf{A}_l}{\partial_t S + q_l V_l}. \qquad (4.84)$$

Chapter 5

Invariantive Mechanics of Systems
of Material Points

5.1. Euclidean invariants and the principle of motion

In Chapter 2 the inertial motion of a material point has been studied. Naturally, arises the problem of studying a system of two material points. Such systems are, for example, the systems composed by two stars (double-star system), or the Solar System. The main property of these systems is the fact that their dimensions are smaller than their mutual distances, and their trajectories are closed to spherical form.

As we shall see, the study of two material points implies existence of two interactions: one attractive, and the other one repulsive ([3], [5], [6], [7], [11]).

Postulate 5.1. *Space and time are structured as variety $E_3 \times T$*

This means that the inertial system of two material points is characterized by the radius-vectors $\mathbf{r}_1$ and $\mathbf{r}_2$, and by the linear momenta $\mathbf{p}_1$ and $\mathbf{p}_2$. In this case, the Euclidean invariants [i.e. quantities which do not modify under Euclidean space transformations (orthogonal transformations)] are

$$\alpha_1 = \frac{1}{2}\mathbf{p}_1^2; \quad \alpha_2 = \frac{1}{2}\mathbf{p}_2^2; \quad \alpha = \mathbf{p}_1 \cdot \mathbf{p}_2;$$

$$\beta_1 = \mathbf{r} \cdot \mathbf{p}_1; \quad \beta_2 = \mathbf{r} \cdot \mathbf{p}_2; \quad \beta = \frac{1}{2}\mathbf{r}^2, \tag{5.1}$$

49

where $\mathbf{r} = \mathbf{r}_2 - \mathbf{r}_1$. The Hamiltonian of the system is a function of the Euclidean invariants (5.1): $H = H(\alpha_1, \alpha_2, \alpha, \beta_1, \beta_2, \beta)$, while the motion is described by the 1-form

$$\omega^i = \sum_{i=1}^{2} \mathbf{p}_i \, d\mathbf{r}_i - H \, dt. \tag{5.2}$$

Postulate 5.2. *The inertial motion of the system of two material points is described by closeness of 1-form (5.2), that is*

$$D\omega^i = \sum_{i=1}^{2} \mathbf{p}_i \wedge d\mathbf{r}_i - dH \wedge dt = 0, \tag{5.3}$$

or, explicitly,

$$D\omega^i = d\mathbf{p}_1 \delta\mathbf{r}_1 - \delta\mathbf{p}_1 d\mathbf{r}_1 + d\mathbf{p}_2 \delta\mathbf{r}_2 - \delta\mathbf{p}_2 d\mathbf{r}_2 - dH\delta t + \delta H dt = 0. \tag{5.4}$$

Since

$$\delta H = \frac{\partial H}{\partial \alpha_1}\mathbf{p}_1\delta\mathbf{p}_1 + \frac{\partial H}{\partial \alpha_2}\mathbf{p}_2\delta\mathbf{p}_2 + \frac{\partial H}{\partial \alpha}(\mathbf{p}_1\delta\mathbf{p}_2 + \mathbf{p}_2\delta\mathbf{p}_1)$$

$$+ \frac{\partial H}{\partial \beta_1}(\mathbf{r}\delta\mathbf{p}_1 + \mathbf{p}_1\delta\mathbf{r}) + \frac{\partial H}{\partial \beta_2}(\mathbf{r}\delta\mathbf{p}_2 + \mathbf{p}_2\delta\mathbf{r}) + \frac{\partial H}{\partial \beta}\mathbf{r}\delta\mathbf{r}, \tag{5.5}$$

by grouping the terms in (5.4), we are left with

$$\left(\frac{d\mathbf{p}_1}{dt} - \frac{\partial H}{\partial \beta_1}\mathbf{p}_1 - \frac{\partial H}{\partial \beta_2}\mathbf{p}_2 - \frac{\partial H}{\partial \beta}\mathbf{r}\right)\delta\mathbf{r}_1$$

$$+ \left(\frac{d\mathbf{p}_2}{dt} + \frac{\partial H}{\partial \beta_1}\mathbf{p}_1 + \frac{\partial H}{\partial \beta_2}\mathbf{p}_2 + \frac{\partial H}{\partial \beta}\mathbf{r}\right)\delta\mathbf{r}_2$$

$$+ \left(-\frac{d\mathbf{r}_1}{dt} + \frac{\partial H}{\partial \alpha_1}\mathbf{p}_1 + \frac{\partial H}{\partial \alpha}\mathbf{p}_2 + \frac{\partial H}{\partial \beta_1}\mathbf{r}\right)\delta\mathbf{p}_1$$

$$+ \left(-\frac{d\mathbf{r}_2}{dt} + \frac{\partial H}{\partial \alpha_2}\mathbf{p}_2 + \frac{\partial H}{\partial \alpha}\mathbf{p}_1 + \frac{\partial H}{\partial \beta_2}\mathbf{r}\right)\delta\mathbf{p}_2 - \frac{dH}{dt}\delta t = 0. \tag{5.6}$$

Multiplying (5.6) by $(dt)^{-1}$ and equating to zero the coefficients of the arbitrary variations $\delta\mathbf{r}_1$, $\delta\mathbf{r}_2$, $\delta\mathbf{p}_1$, $\delta\mathbf{p}_2$, δt, we obtain:

a) The equations of motion

$$\frac{d\mathbf{p}_1}{dt} = \frac{\partial H}{\partial\beta_1}\mathbf{p}_1 + \frac{\partial H}{\partial\beta_2}\mathbf{p}_2 + \frac{\partial H}{\partial\beta}\mathbf{r},$$
$$\frac{d\mathbf{p}_2}{dt} = -\frac{\partial H}{\partial\beta_1}\mathbf{p}_1 - \frac{\partial H}{\partial\beta_2}\mathbf{p}_2 - \frac{\partial H}{\partial\beta}\mathbf{r};$$

(5.7)

b) Definition of linear momenta

$$\mathbf{v}_1 = \frac{\partial H}{\partial\alpha_1}\mathbf{p}_1 + \frac{\partial H}{\partial\alpha}\mathbf{p}_2 + \frac{\partial H}{\partial\beta_1}\mathbf{r};$$
$$\mathbf{v}_2 = \frac{\partial H}{\partial\alpha_2}\mathbf{p}_2 + \frac{\partial H}{\partial\alpha}\mathbf{p}_1 + \frac{\partial H}{\partial\beta_2}\mathbf{r};$$

(5.8)

with

$$\mathbf{v}_1 = \frac{d\mathbf{r}_1}{dt}; \quad \mathbf{v}_2 = \frac{d\mathbf{r}_2}{dt};$$

c) The energy conservation law

$$\frac{dH}{dt} = 0.$$

(5.9)

5.2. Conservation laws

The momentum conservation law

$$\frac{d}{dt}(\mathbf{p}_1 + \mathbf{p}_2) = 0$$

(5.10)

is obtained by adding the two relations (5.7), while the closeness of 1-form (5.2) leads to the energy conservation law (5.9).

The angular momentum $\mathbf{L}$ for the system of two material points is

$$\mathbf{L} = \mathbf{r}_1 \times \mathbf{p}_1 + \mathbf{r}_2 \times \mathbf{p}_2,$$

(5.11)

and, by taking the time derivative

$$\frac{d\mathbf{L}}{dt} = \dot{\mathbf{r}}_1 \times \mathbf{p}_1 + \mathbf{r}_1 \times \dot{\mathbf{p}}_1 + \dot{\mathbf{r}}_2 \times \mathbf{p}_2 + \mathbf{r}_2 \times \dot{\mathbf{p}}_2. \tag{5.12}$$

In view of (5.7) and (5.8), one obtains:

$$
\begin{aligned}
\frac{d\mathbf{L}}{dt} &= \left(\frac{\partial H}{\partial \alpha_1}\mathbf{p_1} + \frac{\partial H}{\partial \alpha}\mathbf{p_2} + \frac{\partial H}{\partial \beta_1}\mathbf{r} \right) \times \mathbf{p_1} \\
&\quad + \mathbf{r}_1 \times \left(\frac{\partial H}{\partial \beta_1}\mathbf{p_1} + \frac{\partial H}{\partial \beta_2}\mathbf{p_2} + \frac{\partial H}{\partial \beta}\mathbf{r} \right) \\
&\quad + \left(\frac{\partial H}{\partial \alpha_2}\mathbf{p_2} + \frac{\partial H}{\partial \alpha}\mathbf{p_1} + \frac{\partial H}{\partial \beta_2}\mathbf{r} \right) \times \mathbf{p_2} \\
&\quad + \mathbf{r}_2 \times \left(-\frac{\partial H}{\partial \beta_1}\mathbf{p_1} - \frac{\partial H}{\partial \beta_2}\mathbf{p_2} - \frac{\partial H}{\partial \beta}\mathbf{r} \right) \\
&= \frac{\partial H}{\partial \beta_1}\mathbf{r} \times \mathbf{p_1} + \frac{\partial H}{\partial \beta_2}\mathbf{r} \times \mathbf{p_2} \\
&\quad - \frac{\partial H}{\partial \beta_1}\mathbf{r} \times \mathbf{p_1} - \frac{\partial H}{\partial \beta_2}\mathbf{r} \times \mathbf{p_2} = 0, \tag{5.13}
\end{aligned}
$$

which is the law of conservation of angular momentum.

5.3. Expressions for linear momenta

Using condition

$$\Delta = \frac{\partial H}{\partial \alpha_1}\frac{\partial H}{\partial \alpha_2} - \left(\frac{\partial H}{\partial \alpha} \right)^2 \neq 0 \tag{5.14}$$

in (5.8), one obtains:

$$\mathbf{p}_1 = \Delta^{-1}\frac{\partial H}{\partial \alpha_2}\mathbf{v}_1 - \Delta^{-1}\frac{\partial H}{\partial \alpha}\mathbf{v}_2 + \Delta^{-1}\left(\frac{\partial H}{\partial \beta_2}\frac{\partial H}{\partial \alpha} - \frac{\partial H}{\partial \alpha_2}\frac{\partial H}{\partial \beta_1} \right)\mathbf{r},$$
$$\tag{5.15}$$
$$\mathbf{p}_2 = \Delta^{-1}\frac{\partial H}{\partial \alpha_1}\mathbf{v}_2 - \Delta^{-1}\frac{\partial H}{\partial \alpha}\mathbf{v}_1 + \Delta^{-1}\left(\frac{\partial H}{\partial \beta_1}\frac{\partial H}{\partial \alpha} - \frac{\partial H}{\partial \alpha_1}\frac{\partial H}{\partial \beta_2} \right)\mathbf{r}.$$

Denoting

$$m_1 = \Delta^{-1}\frac{\partial H}{\partial \alpha_2}; \quad m_2 = \Delta^{-1}\frac{\partial H}{\partial \alpha_1} \tag{5.16}$$

$$\mu = \Delta^{-1}\frac{\partial H}{\partial \alpha}, \tag{5.17}$$

$$\ell_1 = \Delta^{-1}\left(\frac{\partial H}{\partial \beta_2}\frac{\partial H}{\partial \alpha} - \frac{\partial H}{\partial \alpha_2}\frac{\partial H}{\partial \beta_1}\right) = h_1\kappa,$$

$$\ell_2 = \Delta^{-1}\left(\frac{\partial H}{\partial \beta_1}\frac{\partial H}{\partial \alpha} - \frac{\partial H}{\partial \alpha_1}\frac{\partial H}{\partial \beta_2}\right) = h_2\kappa, \tag{5.18}$$

relations (5.15) become

$$\mathbf{p}_1 = m_1\mathbf{v}_1 + \mu\mathbf{v}_2 + h_1\kappa\mathbf{r};$$

$$\mathbf{p}_2 = m_2\mathbf{v}_2 + \mu\mathbf{v}_1 + h_2\kappa\mathbf{r}, \tag{5.19}$$

where m_i $(i = 1, 2)$ are the inertial masses, μ is the gravitational interaction mass, and κ the dilatational interaction mass. Relations (5.19) define linear momenta of the two material points system in invariantive mechanics.

5.4. Expression for energy

If the energy H of the system of two material points is conceived as proper energy corresponding to the three mass species defined by (5.16), (5.17), and (5.18), then

Postulate 5.3. *For the system of two material points, the ratio H/M, with*

$$M = m_1 + m_2 + 2\mu + 2\kappa \tag{5.20}$$

is a universal constant, i.e.

$$\frac{H}{M} = \omega^2. \tag{5.21}$$

According to discussion of Chapter 2, we shall choose $\omega^2 = c^2$, where c is the speed of light in vacuum. The energy of the two-points system therefore is

$$H = (m_1 + m_2 + 2\mu + 2\kappa)c^2, \qquad (5.22)$$

Since

$$m_i = (m_{0i}^2 + m_i \mathbf{v}_i^2 c^{-2})^{1/2}; \quad (i = 1, 2) \qquad (5.23)$$

variation $\delta(c^2 m_i)$ becomes

$$\delta(c^2 m_i) = \mathbf{v}_i \delta(m_i \mathbf{v}_i), \qquad (5.24)$$

or, in view of (5.19) and using

$$\delta F = \delta_1 F + \delta_2 F, \qquad (5.25)$$

where $\delta_1 F$ denotes variation of F with respect to r and $\delta_2 F$ variation of F with respect to the rest of arguments. We can write

$$\delta(c^2 m_1) = \mathbf{v}_1 \delta \mathbf{p}_1 - \mathbf{v}_1 \cdot \mathbf{v}_2 \frac{\partial \mu}{\partial r} \frac{\mathbf{r}}{r} \delta \mathbf{r} - \mathbf{v}_1 \cdot \mathbf{v}_2 \delta_2 \mu$$

$$- \mu \mathbf{v}_1 \cdot \delta \mathbf{v}_2 - \mathbf{v}_1 \cdot \mathbf{r} \frac{\partial \ell_1}{\partial r} \frac{\mathbf{r}}{r} \delta \mathbf{r} - \mathbf{v}_1 \cdot \mathbf{r} \delta_2 \ell_1 - \ell_1 \mathbf{v}_1 \delta \mathbf{r};$$

$$\delta(c^2 m_2) = \mathbf{v}_2 \delta \mathbf{p}_2 - \mathbf{v}_1 \cdot \mathbf{v}_2 \frac{\partial \mu}{\partial r} \frac{\mathbf{r}}{r} \delta \mathbf{r} - \mathbf{v}_1 \cdot \mathbf{v}_2 \delta_2 \mu$$

$$- \mu \mathbf{v}_2 \cdot \delta \mathbf{v}_1 - \mathbf{v}_2 \cdot \mathbf{r} \frac{\partial \ell_2}{\partial r} \frac{\mathbf{r}}{r} \delta \mathbf{r} - \mathbf{v}_2 \cdot \mathbf{r} \delta_2 \ell_2 - \ell_2 \mathbf{v}_2 \delta \mathbf{r}.$$

$$\qquad (5.26)$$

Similarly, one obtains:

$$\delta(c^2 \mu) = c^2 \frac{\partial \mu}{\partial r} \frac{\mathbf{r}}{r} \delta \mathbf{r} + c^2 \delta_2 \mu;$$

$$\qquad (5.27)$$

$$\delta(c^2 \kappa) = c^2 \frac{\partial \kappa}{\partial r} \frac{\mathbf{r}}{r} \delta \mathbf{r} + c^2 \delta_2 \kappa.$$

Variation of H then becomes:

$$\delta H = \mathbf{v}_1 \delta \mathbf{p}_1 + \mathbf{v}_2 \delta \mathbf{p}_2 - \left[\left(2\mathbf{v}_1\mathbf{v}_2 \frac{\partial \mu}{\partial r} - 2c^2 \frac{\partial \mu}{\partial r} - 2c^2 \frac{\partial \kappa}{\partial r} + \mathbf{v}_1 \mathbf{r} \frac{\partial \ell_1}{\partial r} \right. \right.$$

$$\left. \left. + \mathbf{v}_2 \mathbf{r} \frac{\partial \ell_2}{\partial r} \right) \frac{\mathbf{r}}{r} + \ell_1 \mathbf{v}_1 + \ell_2 \mathbf{v}_2 \right] \delta \mathbf{r} + (2c^2 - 2\mathbf{v}_1\mathbf{v}_2)\delta_2 \mu$$

$$- \mu\delta(\mathbf{v}_1\mathbf{v}_2) - \mathbf{v}_1\mathbf{r}\delta_2\ell_1 - \mathbf{v}_2\mathbf{r}\delta_2\ell_2 + 2c^2\delta_2\kappa. \tag{5.28}$$

Substituting (5.28) into (5.4), grouping the terms, and multiplying by $(dt)^{-1}$, we have:

$$\left(\frac{d\mathbf{p}_1}{dt} - \mathbf{L} \right) \delta\mathbf{r}_1 + \left(\frac{d\mathbf{p}_2}{dt} - \mathbf{L} \right) \delta\mathbf{r}_2 + 2c^2 \left(1 - \frac{\mathbf{v}_1\mathbf{v}_2}{c^2} \right) \delta_2\mu$$

$$- \mu\delta(\mathbf{v}_1\mathbf{v}_2) + 2c^2\delta_2\kappa - \mathbf{v}_1\mathbf{r}\delta_2\ell_1 - \mathbf{v}_2\mathbf{r}\delta_2\ell_2 - \frac{dH}{dt}\delta t = 0, \tag{5.29}$$

where

$$\mathbf{L} = \left[2c^2 \left(1 - \frac{\mathbf{v}_1\mathbf{v}_2}{c^2} \right) \frac{\partial \mu}{\partial r} + 2c^2 \frac{\partial \kappa}{\partial r} - \mathbf{v}_1\mathbf{r}\frac{\partial \ell_1}{\partial r} - \mathbf{v}_2\mathbf{r}\frac{\partial \ell_2}{\partial r} \right] \frac{\mathbf{r}}{r}$$

$$- \ell_1 \mathbf{v}_1 - \ell_2 \mathbf{v}_2. \tag{5.30}$$

Equating to zero the coefficients of the arbitrary variations $\delta\mathbf{r}_1$, $\delta\mathbf{r}_2$, and δt, one obtains the equations of motion

$$\frac{d\mathbf{p}_1}{dt} = \mathbf{L}; \quad \frac{d\mathbf{p}_2}{dt} = -\mathbf{L}, \tag{5.31}$$

and the energy conservation law (5.9) as well.

5.5. The inertial masses

5.5.1. *Mass of gravitational interaction*

Let us separate in (5.29) the terms containing μ. The result is

$$2c^2 \left(1 - \frac{\mathbf{v}_1 \cdot \mathbf{v}_2}{c^2} \right) \delta_2\mu = \mu\delta_2(\mathbf{v}_1 \cdot \mathbf{v}_2), \tag{5.32}$$

On the r.h.s., since r does not explicitly appear between parentheses, δ_1 has been replaced by δ_2. Integrating, one obtains

$$\mu = \left(1 - \frac{\mathbf{v_1} \cdot \mathbf{v_2}}{c^2}\right)^{-1/2} \varphi(r), \tag{5.33}$$

where $\varphi(r)$ is an arbitrary function. Its expression is determined under the following conditions:

i) In the limit $v \ll c$, the invariantive theory of the two-particle system must reduce to the Newtonian theory of gravitation. As a result, $\varphi(r)$ and $\mu(r)$ should be

$$\varphi(r) = -\frac{1}{2} \frac{Gm_1^0 m_2^0}{c^2 r}, \tag{5.34}$$

$$\mu = -\frac{1}{2} \frac{Gm_1^0 m_2^0}{c^2 r}, \tag{5.35}$$

where G is Newton's constant.

ii) The invariantive theory of the two particle inertial system must verify the fundamental tests of the gravitational theory: advance of planetary perihelion, deflection of light in Sun's gravitational field, red displacement, radar signal delay, etc.

Analysis of the perihelion advance in invariantive mechanics led to the following expression for $\varphi(r)$ [23], [24]:

$$\varphi(r) = -\frac{1}{2} \frac{Gm_1^0 m_2^0}{c^2 r} \left[1 + \frac{5}{2} \frac{G(m_1^0 + m_2^0)}{c^2 r}\right], \tag{5.36}$$

and for the interaction mass μ

$$\mu = -\frac{1}{2} \frac{Gm_1^0 m_2^0}{c^2 r} \left[1 + \frac{5}{2} \frac{G(m_1^0 + m_2^0)}{c^2 r}\right] \left(1 - \frac{\mathbf{v_1 v_2}}{c^2}\right)^{-1/2}. \tag{5.37}$$

This formula shall be deduced in § 5.9.

5.5.2. *Determination of dilatational interaction mass*

With μ determined, condition (5.9) implies

$$2c^2\delta_2\kappa - \mathbf{v}_1\mathbf{r}\delta_2\ell_1 - \mathbf{v}_2\mathbf{r}\delta_2\ell_2 = 0. \tag{5.38}$$

Since

$$\delta_2\ell_1 = h_1\delta_2\kappa + \kappa\delta_2 h_1;$$
$$\delta_2\ell_2 = h_2\delta_2\kappa + \kappa\delta_2 h_2, \tag{5.39}$$

relation (5.38) becomes

$$\left(1 - \frac{h_1\mathbf{v}_1\mathbf{r} + h_2\mathbf{v}_2\mathbf{r}}{2c^2}\right)\frac{\delta_2\kappa}{\kappa} = \frac{\mathbf{v}_1\mathbf{r}\delta_2 h_1 + \mathbf{v}_2\mathbf{r}\delta_2 h_2}{2c^2}. \tag{5.40}$$

Searching for solution of the form

$$\kappa = \Psi(r)\left(1 - \frac{h_1\mathbf{v}_1\mathbf{r} + h_2\mathbf{v}_2\mathbf{r}}{2c^2}\right)^{-1/2}, \tag{5.41}$$

one finds

$$h_1 = \Psi_1(r)\mathbf{v}_1\mathbf{r}; \quad h_2 = \Psi_2(r)\mathbf{v}_2\mathbf{r}, \tag{5.42}$$

with

$$\Psi_1(r) = \frac{m_1^0}{m_1^0 + m_2^0}\frac{1}{r^2}; \quad \Psi_2(r) = \frac{m_2^0}{m_1^0 + m_2^0}\frac{1}{r^2}. \tag{5.43}$$

Therefore,

$$\kappa = \Psi(r)(1 - u)^{-1/2}, \tag{5.44}$$

where $\Psi(r)$ is an arbitrary function, and

$$u = \frac{m_1^0 v_1^2 \cos^2(\mathbf{v}_1, \mathbf{r}) + m_2^0 v_2^2 \cos^2(\mathbf{v}_2, \mathbf{r})}{2(m_1^0 + m_2^0)c^2}. \tag{5.45}$$

The explicit form of $\Psi(r)$ was given by Onicescu [3], [5], on the basis of of the following assumptions:

i) For infragalactic distances, $\kappa \to 0$;
ii) κ is non-zero for minimum values of r.

Under these conditions,

$$\Psi(r) = -\frac{1}{2}g\ell_{12}\left(\frac{r}{c^2}\right)^2, \tag{5.46}$$

so that the dilatational interaction mass is

$$\kappa = -\frac{1}{2}g\ell_{12}\left(\frac{r}{c^2}\right)^2(1-u)^{-1/2}, \tag{5.47}$$

Here g is a universal constant, and ℓ_{12} a characteristic quantity associated to the two material points of the system.

5.6. The two-body problem in Newtonian mechanics

For a better description of motion of the two bodies within invariantive mechanics, let us first investigate the motion in the frame of Newtonian mechanics.

The bodies are considered as material points, moving under the influence of gravity. The system is considered isolated, *i.e.* the gravitational interaction with other bodies is neglected. Under these assumptions, the centre of mass moves rectilinearly and uniformly relative to a fixed reference frame, while the energy, linear momentum and angular momentum are conserved. In such a frame, the motion is described by the equations

$$m_1^0\ddot{\mathbf{r}}_1 = -\frac{Gm_1^0 m_2^0}{r^2}\frac{\mathbf{r}}{r};$$

$$m_2^0\ddot{\mathbf{r}}_2 = \frac{Gm_1^0 m_2^0}{r^2}\frac{\mathbf{r}}{r}, \tag{5.48}$$

where G is the gravitational constant, m_i^0 $(i = 1, 2)$ are the rest masses, and r the distance between the bodies.

By means of these equations, the equation of motion in a frame of invariable orientation and origin at one of the bodies can be obtained. Consider, for example, the relative motion of the body with mass m_2^0. One finds

$$\ddot{\mathbf{r}} = -\frac{G(m_1^0 + m_2^0)}{r^2}\frac{\mathbf{r}}{r} = -\frac{\mu}{r^2}\frac{\mathbf{r}}{r}, \tag{5.49}$$

where $\mu = G(m_1^0 + m_2^0)$. If $\mathbf{r} \times \dot{\mathbf{r}}$ is the angular momentum taken for unity of mass, we have

$$\frac{d}{dt}(\mathbf{r} \times \dot{\mathbf{r}}) = 0, \tag{5.50}$$

which implies

$$\mathbf{r} \times \dot{\mathbf{r}} = \mathbf{C}. \tag{5.51}$$

Let $(\boldsymbol{\rho}, \boldsymbol{\varepsilon})$, with $\boldsymbol{\rho} = \frac{\mathbf{r}}{r}$, be an orthogonal basis chosen in the plane of motion. In polar coordinates

$$\dot{\mathbf{r}} = \dot{r}\boldsymbol{\rho} + r^2\dot{\theta}\boldsymbol{\varepsilon},$$

so that

$$C = r^2\dot{\theta}. \tag{5.52}$$

On the other hand, if $\frac{1}{2}\dot{\mathbf{r}}^2$ is the kinetic energy per unit mass, then

$$\frac{d}{dt}\left(\frac{1}{2}\dot{\mathbf{r}}^2\right) = \dot{\mathbf{r}}\ddot{\mathbf{r}} = -\frac{\mu}{r^2}\dot{r} = \frac{d}{dt}\left(\frac{\mu}{r}\right), \tag{5.53}$$

therefore

$$\frac{d}{dt}\left(\frac{1}{2}\dot{\mathbf{r}}^2 - \frac{\mu}{r}\right) = 0, \tag{5.54}$$

or

$$\frac{1}{2}\dot{\mathbf{r}}^2 - \frac{\mu}{r} = \frac{h}{2}. \tag{5.55}$$

To determine the trajectory equation in polar coordinates $r = r(\theta)$ we shall use the first integrals (5.52) and (5.55). Using

$$\dot{r}^2 + r^2\dot{\theta}^2 = \frac{2\mu}{r} + h, \tag{5.56}$$

as well as $r^2\dot{\theta} = C$, and eliminating time by means of $r = r(\theta(t))$, we are left with

$$\left[\frac{d}{d\theta}\left(\frac{C}{r}\right)\right]^2 = \frac{2\mu}{r} + h - \frac{C^2}{r^2} = \frac{\mu^2}{C^2} + h - \left(\frac{C}{r} - \frac{\mu}{C}\right)^2. \qquad (5.57)$$

Separating variables

$$d\theta = \pm\frac{d\left(\frac{C}{r} - \frac{\mu}{C}\right)}{\sqrt{\frac{\mu^2}{C^2} + h - \left(\frac{C}{r} - \frac{\mu}{C}\right)^2}}, \qquad (5.58)$$

and integrating, we obtain

$$r = \frac{p}{1 + e\cos\nu}, \qquad (5.59)$$

where

$$p = \frac{C^2}{\mu}; \quad e = \sqrt{1 + \frac{C^2}{\mu^2}h}; \quad \nu = \theta - \theta_0. \qquad (5.60)$$

This is the equation in polar coordinates of a conic section with focus at the origin O, radius vector r, and argument θ with respect to an axis making an angle θ_0 relative to the axis of symmetry of the conic section. If $h < 0$, then $e < 1$ and the conic is an ellipse; if $h = 0$, this yields $e = 1$, and the conic is a parabola; finally, if $h > 0$, this means $e > 1$, corresponding to a hyperbola.

On the other hand, by means of (5.56), (5.59) and (5.60), one obtains

$$v^2 = \mu\left(\frac{2}{r} - \frac{1 - e^2}{p}\right) \qquad (5.61)$$

which leads to

$$h = v^2 - \frac{2\mu}{r} = -\frac{\mu(1 - e^2)}{p}. \qquad (5.62)$$

Denoting $\alpha = \frac{p}{e^2 - 1}$, we can write $h = \frac{\mu}{\alpha}$.

A special case of hyperbolic trajectory is offered by the relative motion of a photon in the gravitational field of a massive body, *e.g.* the Sun. Such an investigation is performed taking into account the corpuscular theory of light, according to which the photon mass is not zero.

If the motion is hyperbolic, then at any time we must have $v^2 > \frac{2\mu}{r}$. According to (5.61), we can write

$$v^2 = \mu\left(\frac{2}{r} + \frac{1}{\alpha}\right); \quad h = \frac{\mu}{\alpha}, \tag{5.63}$$

therefore

$$v^2_{min} = \lim_{r \to \infty} v^2 = \frac{\mu}{\alpha};$$

$$v^2_{max} = \mu\left(\frac{2}{r_0} + \frac{1}{\alpha}\right) = v_0^2, \tag{5.64}$$

where r_0 is the distance between focus and pericentre. Therefore, the photon velocity is not a constant, but varies in the interval $\left[\frac{\mu}{\alpha}, v_0\right]$. Condition for pericentre thus is $v_0^2 > \frac{2\mu}{r_0}$, while condition $\frac{\mu}{\alpha} > \frac{2\mu}{v_0}$ is obviously fulfilled for $r \to \infty$. For the pericentre, $r_0 > \frac{2\mu}{v_0^2}$. But in general relativity, the quantity $\frac{2\mu}{v_0^2} = r_g$ with $v_0 = c$, called Schwarzschild radius, is much smaller than geometrical radius. In case of the Sun, for example, taking $v_0 \approx 3 \cdot 10^8 ms^{-1}$, $m_0 = 2 \cdot 10^{30} kg$, one obtains $r_g \approx 3.10^3 m$. Since the Sun radius is $r_0 = 696 \cdot 10^6 m$, it follows that the photon trajectories with respect to the Sun are hyperbolical and, for $r_g < r < r_0$, these hyperbolas intersect the surface of the Sun. In particular, there is a hyperbola tangent to Sun's surface. On the other hand, if $\mathbf{C} = \mathbf{r} \times \dot{\mathbf{r}} = 0$, the motion is oriented along a straight line, normal to the Sun's surface.

To conclude, trajectories of the photon are hyperbolic and, in particular, linear. From the observational astronomy point of view, the most interesting trajectories are those with pericentral distance bigger than the radius of celestial body (or, at least, equal to it).

Let r_0 be the pericentral distance of the orbit with respect to the center S of the body. For $r \gg r_0$, the motion of photon on hyperbola can be approximated by motion on the two asymptotes. Consequently, using the commonly encountered terms in theory of light refraction [18], deviation of a light beam is determined by the angle δ between extension of the incident semi-asymptote and the refracted semi-asymptote. If $r \to \infty$, relation $1 + e \cos \nu = \frac{p}{r}$ yields $\cos \nu_1 = -1/e$. Since $\varphi = \pi - \nu_1$, we have $\delta = \pi - 2\varphi = 2\nu_1 - \pi$. The problem therefore reduces to determination of ν_1, that is of e. To this end, one takes the derivative of (5.59) and, using (5.52) and (5.60), one obtains

$$e \sin \nu = \frac{p \dot{r}}{r^2 \dot{\nu}} = \frac{p \dot{r}}{C}.$$

In view of (5.59), we still have

$$e^2 = \left(\frac{p}{r} - 1 \right)^2 + \frac{p^2 \dot{r}^2}{C^2} = 1 + \frac{p^2}{r^2} - \frac{2p}{R} + \frac{p^2 \dot{r}^2}{C^2}.$$

This means that, for the pericentre ($r = r_0$, $\dot{r}_0 = 0$), the last formula gives

$$e^2 = 1 + \frac{p^2}{r_0^2} - \frac{2p}{r_0}. \tag{5.65}$$

But $p = C^2/\mu$, and $C = r_0^2 \dot{\theta}_0 = r_0 v_0$, so that one finally obtains

$$e^2 = \frac{r_0^2 v_0^4}{\mu^2} - 2 \frac{v_0^2 r_0}{\mu} + 1. \tag{5.66}$$

Keeping only the first term, we have $e = r_0 v_0/\mu$. Since deviation δ is given by

$$\sin \frac{\delta}{2} = -\cos \nu_1 = \frac{1}{e},$$

where e is greater than unity, a good approximation is given by

$$\delta = \frac{2\mu}{r_0^2 v_0}. \tag{5.67}$$

For the Sun, if the perihelic distance r_0 is considered equal to the Sun radius, one obtains $\delta = 0".87$. In case of Jupiter, which is the largest and most massive planet in our solar system, taking r_0 equal to the planet radius $r_0 = 714 \cdot 10^5 m$, and its mass equal to $M_\odot/1050$, one finds $\delta = 0".017$.

5.7. The two-body problem in general relativity

In this paragraph we present the relative motion of the two bodies within the frame of general relativity. The exposure is based on Chazy's monograph dedicated to connection between general relativity and classical mechanics [25]. The bodies are considered as material points, and the influence from other bodies is neglected. As well known, the most massive body modify the geometric properties of the space, which becomes Riemannian. On the other hand, the geometry of the 4-dimensional space-time variety is given by Schwarzschild metric, emerging from Einstein's equations. This metric, in polar coordinates, writes

$$ds^2 = \left(c^2 - \frac{2\mu}{r}\right) dt^2 - \frac{dr^2}{1 - \frac{2\mu}{c^2 r}} - r^2(d\theta^2 + \cos^2\theta d\varphi^2), \qquad (5.68)$$

where c is the speed of light, r the radial coordinate, θ and φ latitude and longitude, respectively, and t the cosmic time.

The motion of a probe in space-time variety takes place on a geodesic, displayed by means of (5.68). This is a universe line characterized by

$$\delta \int_0^s ds = 0. \qquad (5.69)$$

Taking time as independent variable, we can write

$$\delta \int_0^s ds = \delta \int_0^t f(t, r, \theta, \varphi, r', \theta', \varphi') \, dt = 0, \qquad (5.70)$$

where r', θ', φ' stand for time derivatives, and

$$f(t, r, \theta, \varphi, r', \theta', \varphi')$$

$$= \sqrt{c^2 - \frac{2\mu}{r} - \frac{r'^2}{1 - \frac{2\mu}{c^2 r}} - r^2\theta'^2 - r^2\cos^2\theta\varphi'^2}. \qquad (5.71)$$

Condition (5.70) leads to Euler-Lagrange equationsx

$$\frac{\partial f}{\partial r} - \frac{d}{dt}\frac{\partial f}{\partial r'} = 0;$$

$$\frac{\partial f}{\partial \theta} - \frac{d}{dt}\frac{\partial f}{\partial \theta'} = 0; \qquad (5.72)$$

$$\frac{\partial f}{\partial \varphi} - \frac{d}{dt}\frac{\partial f}{\partial \varphi'} = 0,$$

which is a system of three, second-order differential equations. The general solution of the system depends on six constants of integration, in particular on the orbit elements.

As in Newtonian case, one can show that the trajectory of motion lies in a plane. Indeed, since $\varphi(t)$ does not explicitly interfere in f, equation (5.72) admit the first integral

$$\frac{1}{2}\frac{\partial f}{\partial \varphi'} = -\frac{r^2 \cos^2 \theta \, \varphi'}{f} = const., \qquad (5.73)$$

and the above mentioned property follows immediately.

Therefore, the motion occurs in a meridian plane, defined by the attractive centre and the initial velocity. In this plane, the coordinates are r and θ. Condition (5.69) then becomes

$$\delta \int_0^s ds = \int_0^t f \, dt = \int_0^t \sqrt{c^2 - \frac{2\mu}{r} - \frac{r'^2}{1 - \frac{2\mu}{c^2 r}} - r^2\theta'^2} \, dt. \qquad (5.74)$$

According to Euler-Lagrange equations, since f does not depend of θ, there also exists the first integral

$$\frac{r^2\theta'}{f} = \frac{C}{c}, \qquad (5.75)$$

where C is an arbitrary constant.

A third first integral is obtained due to the fact that f does not contain the independent variable t. It can be written as

$$f - r'\frac{\partial f}{\partial r'} - \theta'\frac{\partial f}{\partial \theta'} = A, \qquad (5.76)$$

where A is another arbitrary constant. One obtains

$$f + \frac{\dfrac{r'^2}{1-\frac{2\mu}{c^2 r}} + r^2 \theta'^2}{f} = A, \qquad (5.77)$$

or

$$\frac{c^2 - \dfrac{2\mu}{r}}{f} = A. \qquad (5.78)$$

Replacing f given by (5.78) into (5.75), one obtains

$$r^2 \frac{d\theta}{dt} = \frac{C}{A}\left(c - \frac{2\mu}{cr}\right). \qquad (5.79)$$

On the other hand, formula (5.78) can be written as

$$\frac{r'^2}{1 - \dfrac{2\mu}{c^2 r}} + r^2 \theta'^2 = c^2 - \frac{2\mu}{r} - \frac{1}{A^2}\left(c^2 - \frac{2\mu}{r}\right)^2. \qquad (5.80)$$

The first integrals (5.79) and (5.80) correspond to the angular momentum and energy first integrals of Newtonian mechanics. To obtain the differential equation of trajectory, one eliminates the time between the first integrals (5.79) and (5.80). The result is

$$\left[\frac{d}{d\theta}\left(\frac{1}{r}\right)\right]^2 = \frac{2\mu}{c^2}\frac{1}{r^3} - \frac{1}{r^2} + \frac{2\mu}{C^2 r} + \frac{A^2 - c^2}{C^2}. \qquad (5.81)$$

It can be observed that the r.h.s. is a third degree polynomial of $\frac{1}{r}$, while in Newtonian case the r.h.s. of the similar equation (5.57) is a second degree polynomial in $\frac{1}{r}$.

Denoting $\frac{1}{r} = u$, equation (5.81) writes

$$\left(\frac{du}{d\theta}\right)^2 = P(u), \qquad (5.82)$$

where

$$P(u) = \frac{2\mu}{c^2}u^3 - u^2 + \frac{2\mu}{C^2}u + \frac{A^2 - c^2}{C^2}.$$

One obtains

$$\theta - \theta_0 = \pm \int \frac{du}{\sqrt{P(u)}}, \tag{5.83}$$

where θ_0 is a constant.

As we have seen, in case of Euclidean approach the integration has been performed by means of circular functions. In relativistic case, the integration needs the use of elliptic functions. But integration can also by performed approximately. For the planets of solar system, r is situated within a bounded interval $[r_1, r_2]$, so that $P(u)$ has two simple roots $u = \alpha$ and $u = \beta$, $(\alpha < \beta)$, but is also has a third root γ, much bigger than α and β. One can write:

$$\theta - \theta_0 = \pm \int \frac{du}{\sqrt{\frac{2\mu}{c^2}(u-\alpha)(\beta-u)(\gamma-u)}}.$$

Let $r_1 = a(1-e)$ and $r_2 = a(1+e)$ be the values of r corresponding to the roots $\beta > \alpha$, that is

$$\beta = \frac{1}{a(1-e)}; \quad \alpha = \frac{1}{a(1+e)}, \tag{5.84}$$

where a is some value of $r \in [r_1, r_2]$, and e a number with property $0 \le e < 1$. It follows that the angle $\Delta\theta$ between two radii corresponding to two consecutive passings at perihelion of the planet is

$$\Delta\theta = 2 \int_\alpha^\beta \frac{du}{\sqrt{\frac{2\mu}{c^2}(u-\alpha)(\beta-u)(\gamma-u)}},$$

because β corresponds to a perihelion, and α to the next aphelion.

To calculate the integral, one makes that change of variables

$$u = \frac{1 + e \cos\nu}{a(1-e^2)}. \tag{5.85}$$

For $u = \beta$, one obtains $\nu = 0$, while $u = \alpha$ yields $\nu = \pi$. One finds

$$\Delta\theta = 2 \int_0^\pi \frac{d\nu}{\sqrt{\frac{2\mu}{c^2}\left[\gamma - \frac{1+e\cos\nu}{a(1-e^2)}\right]}}. \tag{5.86}$$

But $\alpha + \beta + \gamma = \frac{c^2}{2\mu}$. so that

$$\gamma = \frac{c^2}{2\mu} - \frac{2}{a(1 - e^2)}$$

and (5.86) writes

$$\Delta\theta = 2 \int_0^\pi \frac{d\nu}{\sqrt{1 - \frac{6\mu}{c^2 a(1-e^2)} - \frac{2\mu e \cos\nu}{c^2 a(1-e^2)}}}.$$

We can approximate the integral using series expansion of the integrand. Keeping only the terms of the first order, we have

$$\Delta\theta = 2 \int_0^\pi \left\{ 1 + \frac{3\mu}{c^2 a(1 - e^2)} - \frac{\mu e \cos\nu}{c^2 a(1 - e^2)} \right\} d\nu,$$

or

$$\Delta\theta = 2\pi \left[1 + \frac{3\mu}{c^2 a(1 - e^2)} \right].$$

Thus, there exists a rotation of the trajectory symmetry axis in the direction of motion. The perihelion advance, for a rotation period, therefore is

$$\Delta\theta = \frac{6\pi\mu}{c^2 a(1 - e^2)}. \tag{5.87}$$

5.8. The motion of light in fields with Schwarzschild metric

The universe lines (four-dimensional trajectories) described by the light in space-time variety are geodesics of this variety, with the property $ds^2 = 0$. Since the Schwarzschild metric

$$ds^2 = \left(c^2 - \frac{2\mu}{r} \right) dt^2 - \frac{dr^2}{1 - \frac{2\mu}{c^2 r}} - r^2(d\theta^2 + \cos^2\theta\, d\varphi^2)$$

$$= \left(c^2 - \frac{2\mu}{r} \right) dt^2 - d\sigma^2 \tag{5.88}$$

is static, it follows that in the three-dimensional space are geodesics of this space associated with squared differential dt^2 obtained from $ds^2 = 0$, that is

$$c^2 dt^2 = \frac{d\sigma^2}{1 - \frac{2\mu}{c^2 r}} = \frac{dr^2}{\left(1 - \frac{2\mu}{c^2 r}\right)^2} + \frac{r^2(d\theta^2 + \cos^2\theta\, d\varphi^2)}{1 - \frac{2\mu}{c^2 r}}. \tag{5.89}$$

The light trajectories are therefore given by the condition

$$\delta \int_0^\sigma \frac{d\sigma}{\sqrt{1 - \frac{2\mu}{c^2 r}}} = \delta \int \sqrt{\frac{dr^2}{\left(1 - \frac{2\mu}{c^2 r}\right)^2} + \frac{r^2(d\theta^2 + \cos^2\theta\, d\varphi^2)}{1 - \frac{2\mu}{c^2 r}}} = 0. \tag{5.90}$$

On the other hand, the motion of the light along the trajectory is given by

$$ct = \int_0^\sigma \frac{d\sigma}{\sqrt{1 - \frac{2\mu}{c^2 r}}}. \tag{5.91}$$

To study the motion, we shall follow the way displayed in Chazy's monograph [25]. First, it can be shown that the trajectory of the light beam is a plane curve. Indeed, taking θ as independent variable, condition (5.90) becomes

$$\delta \int f\, d\theta = \delta \int \sqrt{\frac{r'^2}{\left(1 - \frac{2\mu}{c^2 r}\right)^2} + \frac{r^2(1 + \cos^2\theta\, \varphi'^2)}{1 - \frac{2\mu}{c^2 r}}}\, d\theta = 0.$$

Since f does not contain φ, Euler's equation $\frac{\partial f}{\partial \varphi} - \frac{d}{dt}\frac{\partial f}{\partial \varphi'} = 0$ provides the first integral

$$\frac{1}{2}\frac{\partial f}{\partial \varphi'} = \frac{r^2 \cos^2\theta\, \varphi'}{\left(1 - \frac{2\mu}{c^2 r}\right) f} = const.,$$

which shows that the trajectory of motion is a plane passing through the attractive centre. In this plane, the coordinates r and θ are polar coordinates. Approximating the radial coordinate r with the radius-vector and taking φ as independent variable, the trajectory of the light beam is given by

$$\delta \int_0^\theta g\, d\theta = \delta \int_0^\theta \sqrt{\frac{r'^2}{\left(1 - \frac{2\mu}{c^2 r}\right)^2} + \frac{r^2}{1 - \frac{2\mu}{c^2 r}}}\, d\theta = 0.$$

Since g does not contain the independent variable θ, there exists the first integral $g - r'\frac{\partial g}{\partial r'} = C$, or

$$g - \frac{r'^2}{\left(1 - \frac{2\mu}{c^2 r}\right)^2 g} = C,$$

which can also be written as

$$\left[\frac{d}{d\theta}\left(\frac{1}{r}\right)\right]^2 = \frac{2\mu}{c^2}\frac{1}{r^3} - \frac{1}{r^2} + \frac{1}{C^2}. \tag{5.92}$$

This equation is similar to (5.81), describing the motion of a body with negligible mass, the only difference being that fact that (5.92) contains no terms in $1/r$ to the first power. Denoting $u = 1/r$, we have

$$\left(\frac{du}{d\theta}\right)^2 = P(u) = \frac{2\mu}{c^2}u^3 - u^2 + \frac{1}{C^2}, \tag{5.93}$$

and, by integration

$$\theta - \theta_0 = \pm \int \frac{du}{\sqrt{P(u)}}. \tag{5.94}$$

Performing — as we did in the previous paragraph — an approximative integration in (5.91), one can investigate the trajectory of the light beam and its angle of deflection as well. The radius vector r of the chosen point takes its minimum value r_0 at the pericentre (in particular, perihelion). We then can write $P(r_0) = 0$. On the other hand, the straight line connecting the pericentre with the attractive centre is a symmetry axis for the trajectory. Let us consider this line as a polar axis, and let $\nu = \theta - \theta_0$ be the polar angle. The constant

C can be obtained from $P(r_0) = 0$, so that we still have

$$P(u) = \frac{2\mu}{c^2} u^3 - u^2 + \frac{1}{r^2}\left(1 - \frac{2\mu}{c^2 r_0}\right). \tag{5.95}$$

Equation $P(u) = 0$, written in the form

$$\frac{1}{r^2} - \frac{1}{r_0^2} = \frac{2\mu}{c^2}\left(\frac{1}{r^3} - \frac{1}{r_0^3}\right)$$

implies

$$\frac{1}{r} + \frac{1}{r_0} = \frac{2\mu}{c^2}\left(\frac{1}{r^2} + \frac{1}{r_0 r} + \frac{1}{r_0^2}\right),$$

or

$$\frac{2\mu}{c^2} u^2 + \left(\frac{2\mu}{c^2 r_0} - 1\right) u + \frac{2\mu}{c^2 r_0^2} - \frac{1}{r_0} = 0,$$

where $r_0 = \frac{2\mu}{c^2}$ is the Schwarzschild radius corresponding to the body.

Performing a series expansion of solutions of the equation and keeping only the main terms, we have

$$u_1 = \alpha = -\frac{1}{r_0} + \frac{3}{2}\frac{\mu}{c^2 r_0^2};$$

$$u_2 = \gamma = \frac{c^2}{2\mu} - \frac{3}{2}\frac{\mu}{c^2 r_0^2}.$$

Consequently, solutions of the equation $P(u) = 0$ satisfy the inequalities

$$\alpha < 0 < \beta = \frac{1}{r_0} < \gamma,$$

if r_0 is of order of the body radius. This means that u takes values in the whole interval $(0, \beta)$, that is r increases from 0 to ∞. The light trajectory is therefore an open curve, with the straight line connecting the attractive centre and the pericentre as axis of symmetry. The concavity of the curve is oriented to the attractive centre, because according to (5.93)

$$\frac{d^2}{d\theta^2}\left(\frac{1}{r}\right) + \frac{1}{r} = \frac{3\mu}{c^2}\cdot\frac{1}{r^2}.$$

If $r \to \infty$, angle ν tends to a limit value ν_1 given by

$$\nu_1 = \int_0^\beta \frac{du}{\sqrt{P(u)}} = \int_0^\beta \frac{du}{\sqrt{\frac{1}{c^2} - u^2 + \frac{2\mu}{c^2}u^3}}.$$

As we have seen, the deviation of the light beam is given by

$$\delta = 2\nu_1 - \pi.$$

The change of variable

$$\frac{w}{C} = u\sqrt{1 - \frac{2\mu}{c^2}u}$$

leads to

$$u = \frac{w}{c}\left(1 - \frac{2\mu}{c^2}u\right)^{-1/2}.$$

Performing a series expansion and keeping only the the first-order terms, we have

$$u \approx \frac{w}{C}\left(1 + \frac{\mu}{c^2}u\right) = \frac{w}{C}\left(1 + \frac{\mu}{c^2}\frac{w}{C}\right).$$

The integral then becomes

$$\nu_1 = \int_0^1 \frac{\left(1 + \frac{2\mu}{c^2 C}\right)dw}{\sqrt{1 - w^2}} = \left[\arcsin w - \frac{2\mu}{c^2 C}\sqrt{1 - u^2}\right]\Bigg|_9^1$$

$$= \frac{\pi}{2} + \frac{2\mu}{c^2 C} = \frac{\pi}{2} + \frac{2\mu}{c^2 r_0}.$$

Therefore, one finally obtains

$$\delta = \frac{4\mu}{c^2 r_0}, \tag{5.96}$$

saying that according to relativistic theory the deviation of the light beam is twice the one predicted by Newtonian mechanics.

5.9. The two-body problem in invariantive mechanics. The perihelion advance

The two bodies can be assimilated to two material points, while the system is considered isolated. Using the investigation shown at the beginning of this chapter, the equations of motion in a system considered to be fixed are

$$\frac{d\mathbf{p}_1}{dt} = \mathbf{L};$$

$$\frac{d\mathbf{p}_2}{dt} = -\mathbf{L}, \tag{5.97}$$

where

$$\mathbf{p}_1 = m_1 \mathbf{v}_1 + \mu \mathbf{v}_2 + h_1 \kappa \mathbf{r};$$

$$\mathbf{p}_2 = m_2 \mathbf{v}_2 + \mu \mathbf{v}_1 + h_2 \kappa \mathbf{r} \tag{5.98}$$

are the linear momenta, and

$$\mathbf{L} = \left\{ 2c^2 \left(1 - \frac{\mathbf{v}_1 \mathbf{v}_2}{c^2} \right) \frac{\partial \mu}{\partial r} + 2c^2 \frac{\partial \mu}{\partial r} - \mathbf{v}_1 \mathbf{r} \frac{\partial \ell_1}{\partial r} - \mathbf{v}_2 \mathbf{r} \frac{\partial \ell_2}{\partial r} \right\} \frac{\mathbf{r}}{r}$$

$$- \ell_1 \mathbf{v}_1 - \ell_2 \mathbf{v}_2. \tag{5.99}$$

It can be shown that mass of dilatational interaction for a two-body compound of the solar system (or even of the galaxy) can be neglected [23], [24], so that in the equations of motion we shall consider

$$\mathbf{p}_1 = m_1 \mathbf{v}_1 + \mu \mathbf{v}_2;$$

$$\mathbf{p}_2 = m_2 \mathbf{v}_2 + \mu \mathbf{v}_1;$$

$$\mathbf{L} = 2c^2 \left(1 - \frac{\mathbf{v}_1 \mathbf{v}_2}{c^2} \right) \frac{\partial \mu}{\partial r} \frac{\mathbf{r}}{r},$$

where the mass of gravitational interaction is

$$\mu = \frac{\varphi(r)}{\sqrt{1 - \frac{\mathbf{v}_1 \mathbf{v}_2}{c^2}}}. \tag{5.100}$$

Function $\varphi(r)$ is going to be determined according to the conditions of motion. As it was shown ([23], [24]), $\varphi(r)$ can be taken in the form

$$\varphi(r) = -\frac{1}{2}\frac{Gm_1^0 m_2^0}{c^2 r}\left(1 + \frac{\overline{k}}{c^2 r}\right), \tag{5.101}$$

where $\overline{k}$ is a parameter.

To obtain the equations of relative motion we use a preferential reference frame. To this end, let us consider a point given be the radius vector

$$\boldsymbol{\rho} = \frac{(m_1 + \mu)\mathbf{r}_1 + (m_2 + \mu)\mathbf{r}_2}{m_1 + m_2 + 2\mu}. \tag{5.102}$$

Since the energy $H = c^2(m_1 + m_2 + 2\mu)$ is conserved, it follows that

$$\dot{\boldsymbol{\rho}} = \frac{\mathbf{p}_1 + \mathbf{p}_2}{m_1 + m_2 + 2\mu} + \frac{\dot{m}_2 + \dot{\mu}}{m_1 + m_2 + 2\mu}\mathbf{r}.$$

We shall prove that, if we limit ourselves to the terms proportional to $\mathbf{v}_1^2/c^2$ and $\mathbf{v}_2^2/c^2$, then $\dot{\boldsymbol{\rho}} = const.$ Using (5.97), we have

$$\frac{d}{dt}\left[m_1^0\mathbf{v}_1\left(1 + \frac{1}{2}\frac{\mathbf{v}_1^2}{c^2}\right) + \mu_0\mathbf{v}_2\right]$$
$$= \frac{Gm_1^0 m_2^0}{r^2}\frac{\mathbf{r}}{r}\left(1 - \frac{1}{2}\frac{\mathbf{v}_1\mathbf{v}_2}{c^2}\right)\left(1 + \frac{2\overline{k}}{c^2 r}\right);$$
$$\frac{d}{dt}\left[m_2^0\mathbf{v}_2\left(1 + \frac{1}{2}\frac{\mathbf{v}_2^2}{c^2}\right) + \mu_0\mathbf{v}_1\right] \tag{5.103}$$
$$= -\frac{Gm_1^0 m_2^0}{r^2}\frac{\mathbf{r}}{r}\left(1 - \frac{1}{2}\frac{\mathbf{v}_1\mathbf{v}_2}{c^2}\right)\left(1 + \frac{2\overline{k}}{c^2 r}\right),$$

where

$$\mu_0 = -\frac{1}{2}G\frac{m_1^0 m_2^0}{c^2 r}.$$

Therefore

$$\dot{\boldsymbol{\rho}} = \frac{\mathbf{p}_1 + \mathbf{p}_2}{m_1 + m_2 + 2\mu_0} + \frac{m_2^0}{m_1 + m_2 + 2\mu_0}\left(\frac{\mathbf{v}_1\mathbf{v}_2}{c^2} + \frac{1}{2}\frac{Gm_1^0\dot{r}}{c^2 r^2}\right)\mathbf{r}$$

$$= \frac{\mathbf{p}_1 + \mathbf{p}_2}{m_1 + m_2 + 2\mu_0} + \frac{m_2^0}{m_1 + m_2 + 2\mu_0}\left(-\frac{\mathbf{v}_2}{c^2}\frac{Gm_1^0}{r} + \frac{1}{2}\frac{Gm_1^0}{c^2 r}\frac{\dot{r}}{r}\mathbf{r}\right).$$

Considering the case of almost circular orbits, which is a good approximation for the planetary motion, we can write

$$\dot{\boldsymbol{\rho}} = \frac{\mathbf{p}_1 + \mathbf{p}_2}{m_1 + m_2 + 2\mu_0} + \frac{m_2^0}{m_1 + m_2 + 2\mu_0}\mathbf{v}_2\frac{\dot{\mathbf{r}}^2}{c^2}$$

$$+ \frac{1}{2}\frac{m_2^0}{m_1 + m_2 + 2\mu_0}\dot{r}\frac{\dot{\mathbf{r}}^2}{c^2}\frac{\mathbf{r}}{r}.$$

Since for the planets $\frac{m_2^0}{m_1^0} < \frac{|\dot{\mathbf{r}}|^2}{c^2}$, we can consider

$$\dot{\boldsymbol{\rho}} = \frac{\mathbf{p}_1 + \mathbf{p}_2}{m_1 + m_2 + 2\mu_0} = const.$$

In other words, the frame with origin at the point defined by $\boldsymbol{\rho}$ is, within the limits of our approximation, an inertial frame. This result was obtained in 1974 [26] and allows one to integrate the equations of motion of invariantive mechanics. In this frame one can write

$$(m_1 + \mu_0)\mathbf{r}_1 = -(m_2 + \mu_0)\mathbf{r}_2;$$

$$(m_1 + \mu_0)\dot{\mathbf{r}}_1 = -(m_2 + \mu_0)\dot{\mathbf{r}}_2; \qquad (5.104)$$

$$(m_1 + \mu_0)r_1 = (m_2 + \mu_0)r_2,$$

where $\mathbf{r}_1$ and $\mathbf{r}_2$ are position vectors. The equations of motion then are

$$\frac{d}{dt}\left[\dot{\mathbf{r}}_1\left(1 + \frac{1}{2}\frac{\dot{\mathbf{r}}_1^2}{c^2} + \frac{Gm_1^0}{c^2 r}\right)\right]$$

$$= \frac{Gm_2^0}{r^2}\frac{\mathbf{r}}{r}\left(1 + \frac{1}{2}\frac{m_1^0}{m_2^0}\frac{\dot{\mathbf{r}}_1^2}{c^2}\right)\left(1 + \frac{2\bar{k}}{c^2 r}\right);$$

$$\frac{d}{dt}\left[\dot{\mathbf{r}}_2\left(1+\frac{1}{2}\frac{\dot{\mathbf{r}}_2^2}{c^2}+\frac{Gm_2^0}{c^2r}\right)\right]$$

$$=-\frac{Gm_1^0}{r^2}\frac{\mathbf{r}}{r}\left(1+\frac{1}{2}\frac{m_2^0}{m_1^0}\frac{\dot{\mathbf{r}}_2^2}{c^2}\right)\left(1+\frac{2\overline{k}}{c^2r}\right). \tag{5.105}$$

Subtracting member by member and keeping only the terms of the second order in $|\dot{\mathbf{r}}|/c$, we have:

$$\frac{d}{dt}\left[\dot{\mathbf{r}}_2\left(1+\frac{1}{2}\frac{\dot{\mathbf{r}}_2^2}{c^2}+\frac{Gm_2^0}{c^2r}\right)-\dot{\mathbf{r}}_1\right]=-\frac{G(m_1^0+m_2^0)}{r^2}\frac{\mathbf{r}}{r}\left(1+\frac{2\overline{k}}{c^2r}\right). \tag{5.106}$$

Using relations

$$\dot{\mathbf{r}}=\dot{\mathbf{r}}_1-\dot{\mathbf{r}}_2;\quad \dot{\mathbf{r}}_2=\frac{m_1+\mu_0}{m_1+m_2+2\mu_0}\dot{\mathbf{r}},$$

we can write

$$\dot{\mathbf{r}}_2\frac{\dot{\mathbf{r}}_2}{c^2}=\dot{\mathbf{r}}\frac{\dot{\mathbf{r}}^2}{c^2}\left(\frac{m_1}{m_1+m_2+2\mu_0}\right)^3=\dot{\mathbf{r}}\frac{\dot{\mathbf{r}}^2}{c^2},$$

and equation (5.106) becomes

$$\frac{d}{dt}\left[\dot{\mathbf{r}}\left(1+\frac{1}{2}\frac{\dot{\mathbf{r}}^2}{c^2}\right)\right]=-\frac{G(m_1^0+m_2^0)}{r^2}\frac{\mathbf{r}}{r}\left(1+\frac{2\overline{k}}{c^2r}\right).$$

Taking derivative with respect to time, we have

$$\ddot{\mathbf{r}}\left(1+\frac{1}{2}\frac{v^2}{c^2}\right)+\dot{\mathbf{r}}\frac{\mathbf{v}\dot{\mathbf{v}}}{c^2}=-\frac{\mu^*}{r^2}\frac{\mathbf{r}}{r}\left(1+\frac{2\overline{k}}{c^2r}\right), \tag{5.107}$$

where v is the modulus of relative velocity, and $\mu^*=G(m_1^0+m_2^0)$.

The trajectory lies in a plane. Indeed, in Cartesian coordinates we can write

$$A\ddot{x}+B\dot{x}+Cx=0;$$

$$A\ddot{y}+B\dot{y}+Cy=0;$$

$$A\ddot{z}+B\dot{z}+Cz=0,$$

where A, B, C are non-zero functions of $r, v, \dot{v}$. It follows that the determinant of the system is identically null. But this determinant is the Wronskian of the functions x, y, z. Consequently, there exists a linear and homogeneous relation with constant coefficients $ax + by + cz = 0$, which is the equation of an invariable plane. In this plane one obtains

$$\ddot{x}\left(1 + \frac{1}{2}\frac{v^2}{c^2}\right) + \frac{\dot{x}\ddot{x} + \dot{y}\ddot{y}}{c^2}\dot{x} = -\frac{\mu^* x}{r^3}\left(1 + \frac{2\overline{k}}{c^2 r}\right);$$

$$\ddot{y}\left(1 + \frac{1}{2}\frac{v^2}{c^2}\right) + \frac{\dot{x}\ddot{x} + \dot{y}\ddot{y}}{c^2}\dot{y} = -\frac{\mu^* y}{r^3}\left(1 + \frac{2\overline{k}}{c^2 r}\right). \tag{5.108}$$

Keeping only the terms up to the second power of v/c (the perihelion advance is a second-order effect), we still have

$$\ddot{x} = \frac{\mu^* x}{r^3}\left(1 - \frac{1}{2}\frac{v^2}{c^2} + \frac{2\overline{k}}{c^2 r}\right) + \frac{\mu^* \dot{r}\dot{x}}{c^2 r^2};$$

$$\ddot{y} = -\frac{\mu^* y}{r^3}\left(1 - \frac{1}{2}\frac{v^2}{c^2} + \frac{2\overline{k}}{c^2 r}\right) + \frac{\mu^* \dot{r}\dot{y}}{c^2 r^2}. \tag{5.109}$$

These equations can be written as

$$\ddot{x} = -\frac{\mu^* x}{r^3} + X; \quad \ddot{y} = -\frac{\mu^* y}{r^3} + Y, \tag{5.110}$$

with

$$X = \frac{\mu^* x}{r^3}\left(\frac{2\alpha\mu^*}{c^2 r} - \frac{2\beta v^2}{c^2}\right) + \frac{\mu^* \dot{r}\dot{x}}{c^2 r^2};$$

$$Y = \frac{\mu^* y}{r^3}\left(\frac{2\alpha\mu^*}{c^2 r} - \frac{2\beta v^2}{c^2}\right) + \frac{\mu^* \dot{r}\dot{y}}{c^2 r^2}, \tag{5.111}$$

and $2\alpha\mu^* = -2k$, $2\beta = -1/2$.

As one observes, the first term of each of equations (5.110) corresponds to the Newtonian force between two bodies in relative motion, while the second corresponds to a perturbative force oriented along the radius vector, and a force guided along velocity.

To integrate the system (5.110) we shall use the method of variation of constants [25]. If, at a given moment, one neglects the perturbative force of components (X, Y), the problem reduces to relative motion of two-body system in Newtonian approach. The initial data are position and velocity on the real trajectory given by (5.110), at the given moment. Thus, the real and Newtonian trajectories have a common point and the same tangent. This is why the Newtonian orbit is called *osculating orbit*. Since the osculating moment of time can be arbitrarily chosen, solution of the system is a conic with time-dependent elements given by Lagrange's equations.

To facilitate the solution to our problem, let us consider the elliptic heliocentric frame, and choose the line of nodes, *i.e.* the straight line of intersection between osculating and orbit planes, as an Ox-axis. Since the motion takes place in a fixed plane, the elements giving its orientation, like orbit inclination and longitude of the ascending node, do not vary. Thus, the Lagrangian approach shall consider only the remaining orbit elements. Choosing as independent variable the eccentric anomaly u instead of time t, we can write:

$$\frac{da}{du} = \frac{2(1 - e\cos u)}{n^2 a}\left(X\frac{\partial x}{\partial \ell_0} + Y\frac{\partial y}{\partial \ell_0}\right);$$

$$\frac{de}{du} = -\frac{\sqrt{1-e^2}(1-e\cos u)}{n^2 ae}\left(X\frac{\partial x}{\partial \omega} + Y\frac{\partial y}{\partial \omega}\right)$$

$$-\frac{(1-\sqrt{1-e^2})(1-e\cos u)}{n^2 a^2 e}\left(X\frac{\partial x}{\partial \ell_0} + Y\frac{\partial y}{\partial \ell_0}\right);$$

$$\frac{d\omega}{du} = -\frac{\sqrt{1-e^2}(1-e\cos u)}{n^2 a^2 e}\left(X\frac{\partial x}{\partial e} + Y\frac{\partial y}{\partial e}\right);$$

$$\frac{d\ell_0}{du} = -\frac{2(1-e\cos u)}{n^2 a}\left(X\frac{\partial x}{\partial a} + Y\frac{\partial y}{\partial a}\right) + (1-\sqrt{1-e^2})\frac{\partial e}{\partial u},$$

$$(5.112)$$

where a is the semi-major axis, e the eccentricity, ω the perihelion longitude, ℓ_0 the mean longitude at $t = 0$, and n the mean angular velocity.

The derivatives of coordinates x and y with respect to the orbit elements are obtained by means of the elliptic trajectory formulas:

$$x = a(\cos u - e)\cos\omega - a\sqrt{1 - e^2}\sin u \sin\omega;$$

$$y = a(\cos u - e)\sin\omega + a\sqrt{1 - e^2}\sin u \cos\omega;$$

$$r = a(1 - e\cos\omega);$$

$$u - e\sin u = nt + \ell_0 - \omega,$$

$$\tag{5.113}$$

while the derivatives $\dot{x}$ and $\dot{y}$ appearing in the perturbative function are deduced in the same way, taking as variables the time and eccentric anomaly.

Calculations give, on the one side, $\frac{da}{du} = 0$; $\frac{de}{du} = 0$, which means that, as in relativistic theory, the semi-major axis and the eccentricity e do not vary. On the other side,

$$\frac{d\omega}{du} = \frac{\mu^*\sqrt{1 - e^2}}{c^2 a e^2}\left\{ \frac{2\beta - 2}{1 - e\cos u} + \frac{2\alpha - 2\beta(1 - e^2) - 4\beta + 4}{(1 - e\cos u)^2} \right.$$

$$\left. + \frac{(4\beta - 2\alpha - 2)(1 - e^2)}{(1 - e\cos u)^3} \right\}. \tag{5.114}$$

Integrating from u and $u + 2\pi$, one obtains the increase $\delta\omega$ of the longitude of the pericentre corresponding to a revolution

$$\delta\omega = \frac{2\pi\mu^*}{c^2 a(1 - e^2)}(2\beta - \alpha + 1). \tag{5.115}$$

In a similar way, using equation for ℓ_0, we find:

$$\delta\ell_0 = \delta\omega + \frac{2\pi\mu^*}{c^2 a}\left[2\beta - 1 + 3\frac{1 - 2\beta}{1 - e\cos u} - 3\frac{\alpha - 2\beta + 1}{(1 - e\cos u)^2} \right]. \tag{5.116}$$

Formula obtained for $\delta\omega$ allows one to determine the parameter $\overline{k}$. Thus, determination of the law of attraction in invariantive mechanics is performed by means of the observed motion. In fact, the Newtonian law of attraction was deduced in the same way, starting from the elliptic motion of planets (the dynamics inverse problem).

In 1879, Le Verrier put into evidence the fact that the motion of perihelion of Mercury is not connected to perturbations produced by the other planets. More precisely, taking into account the interactions with the other planets, he obtained by means of Lagrange equations $\frac{d\omega}{dt} = 527$ arc-seconds per century. On the other hand, processing observation data he obtained $\frac{d\omega}{dt} = 565$ arc-seconds per century. This way, Le Verrier put into evidence a perihelion advance, given by the difference between the observed and calculated angular velocities:

$$\left(\frac{d\omega}{dt}\right)_o - \left(\frac{d\omega}{dt}\right)_c = 38 \text{ arc-seconds per century.}$$

Later, in 1895, Newcomb obtained for the same difference the value 41.24 arc-seconds per century. In the already quoted monograph, Chazy makes a comparative analysis between Le Verrier's and Newcomb's works. He also presents the attempts to explain the perihelion advance before emergence of the of general theory of relativity. According to this theory, variation $\delta\omega$ is

$$\delta\omega = \frac{6\pi\mu^*}{c^2 a(1 - e^2)}. \tag{5.117}$$

The advance predicted by the theory of relativity is in harmony with that deduced from observations. It is worthwhile to mention that the pre-relativistic attempts led in general to formula (5.117), except for a factor. In invariantive mechanics one obtains the similar formula (5.115). If $2\beta - \alpha + 1 = 3$, *i.e.* if $\alpha = 5/2$, (5.115) goes to (5.117), which can be considered as being the closest to reality. But α is connected to $\overline{k}$, which interferes in function $\varphi(r)$. In its turn, $\varphi(r)$ appears on the integration leading to determination of the mass of gravitational interaction μ. This way, one obtains

$$\overline{k} = \frac{5}{2}G(m_1^0 + m_2^0),$$

and

$$\varphi(r) = -\frac{1}{2}\frac{Gm_1^0 m_2^0}{c^2 r}\left[1 + \frac{5}{2}\frac{G(m_1^0 + m_2^0)}{c^2 r}\right]. \tag{5.118}$$

5.10. Hubble's law

Consider the linear momentum conservation law

$$m_1\mathbf{v}_1 + m_2\mathbf{v}_2 + \mu(\mathbf{v}_1 + \mathbf{v}_2) + (h_1 + h_2)\kappa\mathbf{r} = \mathbf{C}. \tag{5.119}$$

Recalling expressions of h_1 and h_2, and observing that

$$\mathbf{v}_i\mathbf{r} = v_i r \cos(\mathbf{v}_i, \mathbf{r}) = r\xi_i, \tag{5.120}$$

where

$$\xi_i = v_i \cos(\mathbf{v}_i, \mathbf{r}), \tag{5.121}$$

then multiplying (5.119) by $\frac{\mathbf{r}}{r}$, we have

$$m_1\xi_1 + m_2\xi_2 + \mu(\xi_1 + \xi_2) + \frac{m_1^0\xi_1 + m_2^0\xi_2}{m_1^0 + m_2^0}\frac{1}{r^2}\kappa r^2 = C\cos(\mathbf{C}, \mathbf{r}). \tag{5.122}$$

Dividing this relation by $m_1\xi_1 + m_2\xi_2 - C\cos(\mathbf{C}, \mathbf{r}) \neq 0$ and keeping in mind (5.37) and (5.41), we are left with

$$1 + \frac{1}{c^2 r}M + \frac{1}{c^4}Nr^2 = 0, \tag{5.123}$$

where

$$M = -\frac{1}{2}G\frac{m_1^0 m_2^0}{\left(1 - \frac{\mathbf{v}_1\mathbf{v}_2}{c^2}\right)^{1/2}}$$

$$\times \frac{\xi_1 + \xi_2}{m_1\xi_1 + m_2\xi_2 - C\cos(\mathbf{C}, \mathbf{r})}\left(1 + \frac{5}{2}G\frac{m_1^0 + m_2^0}{c^2 r}\right), \tag{5.124}$$

and

$$N = -\frac{1}{2}G\frac{\ell_{12}}{m_1^0 + m_2^0} \times \frac{m_1^0\xi_1 + m_2^0\xi_2}{m_1\xi_1 + m_2\xi_2 - C\cos(\mathbf{C},\mathbf{r})}\frac{1}{(1-u)^{1/2}}.$$

$$(5.125)$$

Taking the time derivative of (5.123) and dividing by $\left(\frac{2r}{c^4}N\right)$, we have

$$\frac{c^2}{2r^2}\frac{\dot{M}}{N} + \frac{c^2}{2r^3}\frac{M}{N}\frac{\mathbf{r}\dot{\mathbf{r}}}{r} + \frac{\mathbf{r}\dot{\mathbf{r}}}{r} + \frac{1}{2}\frac{\dot{N}}{N}r = 0,$$

$$(5.126)$$

which yields

$$\left(1 - \frac{c^2}{2r^3}\frac{M}{N}\right)\frac{\mathbf{r}\dot{\mathbf{r}}}{r} = -\frac{1}{2}\frac{\dot{N}}{N}r\left(1 + \frac{c^2}{r^3}\frac{\dot{M}}{\dot{N}}\right).$$

$$(5.127)$$

One then obtains

$$|\mathbf{v}_2\cos(\mathbf{v}_2,\mathbf{r}) - \mathbf{v}_1\cos(\mathbf{v}_1,\mathbf{r})| = \theta r,$$

$$(5.128)$$

with

$$\theta = \frac{1}{2}\left|\frac{\dot{N}}{N}\right|\frac{\left|1 + \frac{c^2\dot{M}}{r^2\dot{N}}\right|}{\left|1 - \frac{c^2}{2r^3}\frac{M}{N}\right|}.$$

$$(5.129)$$

Quantity θ can be considered to be approximately constant for large intervals of velocity [5], saying that in invariantive mechanics the linear momentum conservation law manifests itself as Hubble's law.

5.11. Invariantive mechanics and interactions between bodies

According to the results displayed in the previous paragraphs, the interactions between bodies within the frame of invariantive mechanics are of two types. The first type concerns the attractive interaction given by

$$L_a = 2c^2 \left(1 - \frac{\mathbf{v}_1 \mathbf{v}_2}{c^2}\right) \frac{\partial \mu}{\partial r}.$$

In view of (5.100) and (5.101), we still have

$$L_a = \frac{Gm_1^0 m_2^0}{c^2} \left(1 - \frac{\mathbf{v}_1 \mathbf{v}_2}{c^2}\right)^{1/2} \left(1 + \frac{2\overline{k}}{c^2 r}\right)$$

with $\overline{k} = \dfrac{5}{2}G(m_1^0 + m_2^0)$. For $|\mathbf{v}_1| \ll c$, $|\mathbf{v}_2| \ll c$, we can write

$$L_a = \frac{Gm_1^0 m_2^0}{c^2} \left(1 - \frac{1}{2}\frac{\mathbf{v}_1 \mathbf{v}_2}{c^2} + \frac{2\overline{k}}{c^2 r} + \dots\right). \tag{5.130}$$

As one observes, the first term of expansion is precisely the Newtonian force. The attractive interaction is dominant for infragalactic distances.

For bigger distances, the repulsive interaction becomes sensitive. This type of interaction, characterized by the dilatational mass κ, is given by

$$L_r = \frac{g_{12} r}{c^2 \sqrt{1 - u}} + \dots, \tag{5.131}$$

where $g_{12} = 2g\ell_{12}$. Therefore, the force is proportional to the distance. This force is specific to invariantive mechanics and could be used to explain expansion of the spatial volume of the observable Universe (Hubble's law).

5.12. One-body problem in invariantive mechanics

5.12.1. *First integrals and the equation of motion*

The problem of a single body in invariantive mechanics concerns the motion of a probe-body in the gravitational field generated by a source-body [27], [28]. Among the probe-bodies one can enumerate the planets, the comets, the photon, etc., while the Sun is a source-body. Imposing restrictions like

$$m_2^0 \gg m_1^0; \quad \mathbf{v}_2 \equiv 0; \quad \mathbf{v}_1 = \mathbf{v}; \quad \kappa \equiv 0, \tag{5.132}$$

the angular momentum and energy first integrals become

$$r^2\dot{\theta} = L\left(1 - \frac{\mathbf{v}^2}{c^2}\right)^{1/2}; \quad L = const., \tag{5.133}$$

together with

$$m_1 c^2 + 2\mu c^2 = E = const.;$$

$$m_1 = m_0\left(1 - \frac{\mathbf{v}^2}{c^2}\right)^{-1/2}; \tag{5.134}$$

$$\mu = -\frac{Gm_1^0 m_2^0}{c^2 r}\left(1 + \alpha\frac{Gm_2^0}{c^2 r}\right).$$

Using (5.133) and (5.134), we have

$$\frac{\mathbf{v}^2}{c^2} = \frac{\dot{r}^2 + r^2\dot{\theta}^2}{c^2} = \left[\left(\frac{du}{d\theta}\right)^2 + u^2\right]\left(1 - \frac{\mathbf{v}^2}{c^2}\right)\frac{L^2}{c^2}; \quad u = \frac{1}{r}, \tag{5.135}$$

and

$$\left(1 - \frac{\mathbf{v}^2}{c^2}\right) - \frac{1}{\left(\varepsilon - \frac{2\mu}{m_1^0}\right)^2}; \quad \frac{E}{m_1^0 c^2} - \varepsilon. \tag{5.136}$$

Using these observations, we are now able to write the differential equation of motion

$$\left(\frac{du}{d\theta}\right)^2 + u^2 = \frac{c^2}{L^2}\left[\left(\varepsilon - \frac{2\mu}{m_1^0}\right)^2 - 1\right]. \tag{5.137}$$

Replacing μ given by (5.134), we still have

$$\left(\frac{du}{d\theta}\right)^2 = P(u), \tag{5.138}$$

where

$$P(u) = \alpha^2 n^2 \left(\frac{nc}{L}\right)^2 u^4 + 2\alpha n \left(\frac{nc}{L}\right)^2 u^3 + \left[(2\alpha\varepsilon + 1)\left(\frac{nc}{L}\right)^2 - 1\right] u^2$$

$$+ 2\varepsilon n \left(\frac{c}{L}\right)^2 u + (\varepsilon^2 - 1)\left(\frac{c}{L}\right)^2; \quad n = \frac{Gm_2^0}{c^2}. \tag{5.139}$$

Taking the derivative with respect to θ and performing the appropriate calculations, we finally obtain

$$u'' + u\left[1 - \left(\frac{nc}{L}\right)^2 (2\alpha\varepsilon + 1)\right] = \left(\frac{c}{L}\right)^2 (2\alpha^2 n^4 u^3 + 3\alpha n^3 u^2 + n\varepsilon).$$

$$\tag{5.140}$$

The general solution to the problem is

$$\theta - \theta_0 = \pm \int \frac{du}{\sqrt{P(u)}}; \quad \theta_0 = const.$$

This integration is performed by means of elliptic functions. Nevertheless, we can use an approximation procedure, observing that the trajectory of the real motion is a conic section (either an ellipse in case of the perihelion advance, or a hyperbola in case of the photon deflection). To this end, let us evaluate the order of magnitude of each term in $P(u)$. Introducing the characteristic velocity q and the characteristic length l, we have

$$\varepsilon^2 - 1 \sim \frac{q^2}{c^2}; \quad u^2 \sim \frac{1}{l^2}; \quad n \sim l\frac{q^2}{c^2}; \quad u \sim \frac{1}{l}.$$

In view of these approximations, it follows that the terms of zero-th, first and second order with regard to u are of the order of magnitude $1/l^2$, while the terms of third and fourth order in regards to u are of order of magnitude $q^2/c^4 l^2$ and $q^6/c^6 l^2$, respectively. The equation of motion (5.138) then becomes

$$\left(\frac{du}{d\theta}\right)^2 = P(u) = \left(\frac{c}{L}\right)^2 (\varepsilon^2 - 1) + 2\varepsilon n \left(\frac{c}{L}\right)^2 u$$

$$+ \left[(2\alpha\varepsilon + 1)\left(\frac{nc}{L}\right)^2 - 1\right] u^2. \tag{5.141}$$

5.12.2. *The perihelion advance*

For

$$\varepsilon = 1 + \delta < 1; \quad |\delta| \ll 1;$$

$$\delta = \frac{v_0^2}{2c^2} - \frac{n}{r_0}\left(1 + \alpha\frac{n}{r_0}\right) = const.; \tag{5.142}$$

$$|\mathbf{v}_0| = const.; \quad r_0 = const.,$$

the polynomial (5.141) admits two positive roots r_1, r_2 which satisfy condition $r_1 < r < r_2$ (bounded motion of elliptical type). Let us take the roots r_1, r_2 in the form

$$u_1 = \frac{1}{r_1} = \frac{1 + e}{p};$$

$$u_2 = \frac{1}{r_2} = \frac{1 - e}{p}, \tag{5.143}$$

where e and p are some new constants which depend on ε and L. We then have

$$1 - \varepsilon^2 = \frac{n}{p}(1 - e^2);$$

$$L^2 = nc^2 p; \tag{5.144}$$

$$\nu^2 = 1 - \frac{(2\alpha\varepsilon + 1)}{p},$$

and (5.141) becomes

$$\frac{1}{\nu^2}\left(\frac{du}{d\theta}\right)^2 \approx \frac{e^2 - 1}{p} + \frac{2}{p}u - u^2. \tag{5.145}$$

The solution of this equation is

$$u = \frac{1 + e\cos\nu\theta}{p}. \tag{5.146}$$

The constant of integration in (5.146) has been chosen so that the minimum value of r (maximum value of u) corresponds to $\theta = 0$.

In this case the radius-vector r reverts to its initial value for an increase of θ greater than 2π. The difference

$$\delta\omega = \frac{2\pi}{\nu} - 2\pi = \frac{2\pi\left(\alpha\varepsilon + \frac{1}{2}\right)n}{p} \qquad (5.147)$$

gives the perihelion advance corresponding to a period of revolution of the planet.

The value of α is obtained by identifying (5.147), if approximations (5.142a,b) are considered, with the result given by astronomical observations, which is

$$\frac{2\pi\left(\alpha + \frac{1}{2}\right)n}{p} = \frac{6\pi n}{p}. \qquad (5.148)$$

One reobtains the value $\alpha = 5/2$ ([27], [28]).

5.12.3. *The photon deflection*

The field theories associate the photon with the electromagnetic field, as a particle with null rest mass and velocity c [29]. The formalism presented in "one-body problem" does not allow to conceive the photon as a field particle, due to singularities arising for $v = c$. Nevertheless, the photon problem can be studied in invariantive mechanics, assuming that the photon rest mass is non-zero, and its velocity is $v < c$, with c as an asymptotic limit. This hypothesis is not singular, it was first introduced by Louis de Broglie in his theory on duality wave-corpuscle [21] and, more recently, as field theories with massive photons (Proca, [16]), or in modern cosmological theories concerning the photon aging [30].

Let us first observe that, if

$$\varepsilon = 1 + \delta > 1, \quad \delta \ll 1, \qquad (5.149)$$

one of the roots of polynomial (5.141) (r_1 or r_2) is negative. Denoting by r_1 the positive root, we have $r_1 < r$ and the orbit breaks away to infinity (non-limited motion, of hyperbolic type). Taking the derivative of (5.151), we have

$$u'' + u \approx n \left(\frac{c}{L}\right)^2 + (2\alpha + 1) \left(\frac{cn}{L}\right)^2 u, \qquad (5.150)$$

or, if (5.137) is taken into account,

$$u'^2 + u^2 \approx 2n \left(\frac{c}{L}\right)^2; \qquad (5.151)$$

$$u'' + u \approx \left(\frac{c}{L}\right)^2 + \left(\alpha + \frac{1}{2}\right) n(u'^2 + u^2). \qquad (5.152)$$

The asymptotic condition $v \to c$, in agreement with (5.133), implies

$$L \to \infty; \quad \left|\frac{\dot{r}}{r\dot{\theta}}\right| \to 0, \qquad (5.153)$$

in which case (5.152) becomes

$$u'' + u \approx \left(\alpha + \frac{1}{2}\right) nu^2. \qquad (5.154)$$

To integrate this equation, we shall use the method of successive approximations. Noticing that the effect of gravitational mass is contained in the right member of the equation, the differential equation

$$u'' + u \approx 0 \qquad (5.155)$$

corresponds to the photon trajectory for $\mu \equiv 0$, which corresponds to the absence of gravity. Solution of (5.155) is the straight line

$$\sigma_0 = \frac{\cos \theta}{R}, \qquad (5.156)$$

where by σ_0 has been denoted the zero-th order approximation, and by $1/R$ the constant of integration.

To obtain the approximation of the first-order, one introduces (5.156) into the right member of (5.154). The result is

$$u'' + u = \left(\alpha + \frac{1}{2}\right) n \frac{\cos^2 \theta}{R^2},$$
(5.157)

which admits as solution

$$u_1 = \left(\alpha + \frac{1}{2}\right) \frac{n}{3R^2} (\cos^2 \theta + 2\sin^2 \theta).$$
(5.158)

This result can be easily verified, by taking the derivative. In the first-order of approximation, the solution of equation (5.154) therefore is

$$u = u_0 + u_1 = \frac{\cos \theta}{R} + \left(\alpha + \frac{1}{2}\right) \frac{n}{3R^2} (\cos^2 \theta + 2\sin^2 \theta).$$
(5.159)

This is the finite equation of photon trajectory in the gravitational field of the Sun.

Passing from polar coordinates r, θ to Cartesian coordinates x, y by means of

$$x = r\cos\theta; \quad y = r\sin\theta,$$

solution (5.159) becomes

$$x = R - \left(\alpha + \frac{1}{2}\right) \frac{n}{3R} \frac{x^2 + 2y^2}{(x^2 + y^2)^{1/2}}.$$
(5.160)

Geometrically, this is an arc of hyperbola, with source of the gravitational field (the Sun) in its focus.

The magnitude of deviation is given by angle θ between the tangent to trajectory and the straight line $x = R$, that is

$$\tan \frac{\theta}{2} = \lim_{y\to\infty} \frac{R - x}{y} = \left(\alpha + \frac{1}{2}\right) \frac{n}{3R} \lim_{y\to\infty} \frac{x^2 + 2y^2}{(x^2 + y^2)^{1/2}}$$

$$= \left(\alpha + \frac{1}{2}\right) \frac{2n}{3R}.$$
(5.161)

For small angles,

$$\tan\frac{\theta}{2} \approx \frac{\theta}{2} = \left(\alpha + \frac{1}{2}\right)\frac{2n}{3R}, \qquad (5.162)$$

which gives

$$\theta = \left(\alpha + \frac{1}{2}\right)\frac{4n}{3R}. \qquad (5.163)$$

The magnitude of α is obtained by identifying (5.163) with data given by astronomical observations, that is

$$\left(\alpha + \frac{1}{2}\right)\frac{4n}{3R} = \frac{4n}{R}. \qquad (5.164)$$

One finds $\alpha = 5/2$ [28].

Chapter 6

The Photon in Invariantive Ondulatory Theories

6.1. The red shift

In agreement with results of Chapter 4, the equivalence between the inertial 1-form $\omega^i = \mathbf{p}d\mathbf{r} - Hdt$ and the ondulatory 1-form $\omega^o = \mathbf{k}d\mathbf{r} - \omega dt$ imply the definition relations for linear momentum $\mathbf{p}$ and energy H:

$$\mathbf{p} = \hbar\mathbf{k}; \tag{6.1a}$$

$$H = mc^2 = \hbar\omega, \tag{6.1b}$$

where $\hbar = h/2\pi$ is the reduced Planck constant. According to (6.1b), the photon's inertial mass is

$$m_i = \frac{\hbar\omega}{c^2}. \tag{6.2}$$

On the other hand, following Einstein's equivalence principle [30], the photon must have a field mass m_f, equal to inertial mass $m_i = m_f \equiv m$. In the presence of the source-body, the energy of gravitational interaction $m_f V(r) = -Gm_2^0 m/r$ has to be added to the energy of the free photon $mc^2 = \hbar\omega$. Supposing that the photon coherence length is much smaller than the distance between photon and source-body, and the photon is point-like with a good approximation, the energy conservation law becomes

$$H = \hbar\omega\left(1 - \frac{n}{r}\right) = const. \tag{6.3a}$$

with

$$n = \frac{Gm_2^0}{c^2}. \tag{6.3b}$$

The same result (6.3a) can be obtained by means of the equation of conservation of energy encountered in "one-body problem"

$$H = (m + 2\mu_0)c^2 = const.,$$

if one replaces mc^2 by $\hbar\omega$, and m^0 by $\hbar\omega/c^2$ in mass of gravitational interaction formula

$$\mu_0 = -\frac{Gm_2^0 m^0}{c^2 r}$$

Let ω be the pulsation of electromagnetic radiation emitted by an atom situates at the surface of a star of radius R, and ω' pulsation of that radiation registered at the distance $r' \gg R$, for example at the surface of the Earth. According to (6.3a), we have

$$\hbar\omega \left(1 - \frac{n}{R}\right) = \hbar\omega' \left(1 - \frac{n}{r'}\right),$$

or, in view of approximation $r' \gg R$,

$$\omega' \approx \omega \left(1 - \frac{n}{R}\right) < \omega, \tag{6.4}$$

corresponding to a red shift of some spectral line emitted by the atom. The relative shift

$$\frac{\Delta\lambda}{\lambda} = = \frac{\Delta\omega}{\omega} = \frac{n}{R}$$

depends on the "strength" of the gravitational field.

6.2. The photon deflection

Let us study the interaction photon-gravitational field by means of a Lagrangian procedure. The explicit form of the Lagrangian L is based on the following considerations: i) The kinetic term T and the potential term V are separated in L, that is

$$L = T - V. \tag{6.5}$$

Such a situation is not singular in invariantive mechanics. For example, the classical Lagrangian

$$L = \mathbf{p}_1 \mathbf{v} - (m_1 + 2\mu^0)c^2 \approx \frac{1}{2}m_1^0 \mathbf{v}^2 + \frac{Gm_1^0 m_2^0}{r}$$

satisfies property (6.5). Since T and V are given by [30]

$$T = mc^2 = \hbar\omega; \tag{6.6}$$

$$V = -2\mu^0 c^2 = \frac{\hbar\omega}{c^2}\frac{GM_2^0}{r}, \tag{6.7}$$

the Lagrangian L writes

$$L = (m - 2\mu^0)c^2 = \hbar\omega\left(1 + \frac{n}{r}\right). \tag{6.8}$$

ii) In the presence of a gravitational field, the photon mass modifies according to (see 6.3a)

$$\frac{\hbar\omega}{c^2}\left(1 - \frac{n}{r}\right) = \frac{\hbar\omega_0}{c^2} = const., \tag{6.9}$$

and (6.8) becomes

$$L = \hbar\omega_0 N(r) \tag{6.10a}$$

with

$$N(r) = \left(1 - \frac{n}{r}\right)^{-1}\left(1 + \frac{n}{r}\right). \tag{6.10b}$$

iii) The photon velocity has a constant modulus, but variable orientation. The Lagrangian (6.10a) takes the form

$$L = \frac{\hbar\omega_0}{c}N(r)\,(\dot{r}^2 + r^2\dot{\theta}^2)^{1/2}. \tag{6.11}$$

The Euler-Lagrange equations

$$\frac{d}{dt}\left(\frac{\partial L}{\partial \dot{\theta}}\right) - \frac{\partial L}{\partial \theta} = 0; \tag{6.12a}$$

$$\frac{d}{dt}\left(\frac{\partial L}{\partial \dot{r}}\right) - \frac{\partial L}{\partial r} = 0 \tag{6.12b}$$

with L given by (6.11), then yield

$$N(r)r^2\dot{\theta} = L_0 = cR = const. \tag{6.13a}$$

$$\frac{d}{dt}\left[N(r)\frac{\dot{r}}{c}\right] = c\frac{\partial N(r)}{\partial r} + N(r)\frac{r\dot{\theta}^2}{c}. \tag{6.13b}$$

The equation of motion follows by eliminating the time t between (6.13a) and (6.13b)

$$u'' + u = \frac{N(r)}{r^2}\frac{\partial N(u)}{\partial u}; \quad u = \frac{1}{r}, \tag{6.14}$$

or, explicitly

$$u'' + u = \frac{2n}{R^2}\frac{(1+nu)}{(1-nu)^3}. \tag{6.15}$$

Since $nu \ll 1$, (6.15) can be written as

$$u'' + u \approx \frac{2n}{R^2}. \tag{6.16}$$

This equation admits a solution of the form

$$u = \frac{1 + e\cos\theta}{p}, \tag{6.17}$$

with

$$\frac{1}{p} = \frac{2n}{r^2}; \tag{6.18a}$$

$$\frac{e}{p} = \frac{1}{R}. \tag{6.18b}$$

The integration constant in (6.17) is chosen in such a way that the value $\theta = 0$ corresponds to the maximum value of u (*i.e.* minimum value of r). We approximately have

$$r_{min} = R - 2n.$$

In the Euclidean plane, (6.17) represents a hyperbola, whose asymptotic directions are determined by the condition $u = 0$, which implies

$$\cos\theta = -\frac{2n}{R}. \tag{6.19}$$

The limited values for angle θ are

$$\theta_1 = \frac{\pi}{2} + \delta; \quad \theta_2 = \frac{\pi}{2} - \delta, \tag{6.20}$$

with

$$\delta = \frac{2n}{R}. \tag{6.21}$$

The deviation angle for the light beam is the angle between hyperbola asymptotes, that is

$$\Delta\theta = 2\delta = \frac{4n}{R}, \tag{6.22}$$

in agreement with astronomical observations.

6.3. Fermat's principle. Gravitational retardation of light

Correspondence with the invariantive ondulatory theory needs existence of a Fermat principle. Indeed, since variation of H is null along trajectory — see (6.3a)

$$\delta H = 0, \tag{6.23}$$

the variational principle

$$\delta \int d\omega^i = \delta \int ((\mathbf{pv} - H)\, dt = 0 \tag{6.24}$$

reduces to

$$\delta \int \mathbf{p}\, dr = 0. \tag{6.25}$$

In agreement with Chapter 4, for the photon we have $d\omega^i = 0$, which means $H = \mathbf{pv}$, and (6.25) yields

$$\delta \int \frac{H}{|\mathbf{v}|}|d\mathbf{r}| = \delta \int \frac{H}{C} N(r)|d\mathbf{r}| = 0 \; ; \quad \frac{c}{|\mathbf{v}|} = N(r) \tag{6.26}.$$

Since $H/c = const.$, we are left with Fermat's principle

$$\int N(r)\,|d\mathbf{r}| = 0. \tag{6.27}$$

In this context, the Lagrangian of the photon writes

$$L = const.N(r)|\mathbf{v}| = const.N(r)(\dot{r}^2 + r^2\dot{\theta}^2)^{1/2}. \tag{6.28}$$

For $\hbar\omega_0/c = const.$, relations (6.28), (6.11), and (6.10b) lead to

$$N(r) = \left(1 - \frac{n}{r}\right)^{-1} \left(1 + \frac{n}{r}\right), \tag{6.29}$$

showing that gravitation can be assimilated with a medium with refractive index $N(r)$. In particular, for $n/r \ll 1$, one finds [30]

$$N(r) \approx 1 + \frac{2n}{r}. \tag{6.30}$$

Based on this refractive index, in 1966 Irwin Shapiro (Lincoln Laboratory, Massachusetts Institute of Technology) imagined a new test of the general relativity, as follows. If a light beam passing between two planets goes close to the Sun, then the passing time interval is bigger. Indeed, if the Sun is absent, the time duration of the light beam pass between two specified positions $-l_1$ and $+l_2$ is

$$t_0 = \frac{1}{c} \int_{-l_1}^{+l_2} dy. \tag{6.31}$$

In the Sun's presence, the analogous duration is

$$t_S = \int_{-l_1}^{+l_2} \frac{dy}{v} = \frac{1}{c} \int_{-l_1}^{+l_2} N(r)\, dy. \qquad (6.32)$$

The resulting delay is

$$\Delta t = \frac{1}{c} \int_{-l_1}^{+l_2} [N(r) - 1]\, dy \approx 2n \int_{-l_1}^{+l_2} \frac{dy}{r}. \qquad (6.33)$$

The effect reaches a maximum if the light beam is tangent to the Sun surface. Since $r = (R^2 + y^2)^{1/2}$, where r is the Sun radius, and $l_1 \gg R$, $l_2 \gg R$, the time difference (delay) writes

$$\Delta t = 2\pi \int_{-l_1}^{+l_2} \frac{dy}{(R^2 + y^2)^{1/2}} = 2n \ln \frac{(R^2 + l_2^2)^{1/2} + l_2}{(R^2 + l_1^2)^{1/2} + l_1}$$

$$= 2n \ln \left\{ \left[\frac{l_2}{R} + \left(\frac{l_2^2}{R^2} + 1 \right)^{1/2} \right] \Big/ \left[\frac{l_1}{R} + \left(\frac{l_1^2}{R^2} + 1 \right)^{1/2} \right] \right\}$$

$$\approx 2n \ln \left(4 \frac{l_1 l_2}{R^2} \right). \qquad (6.34)$$

Irvin Shapiro verified the existence of this delay using radar signals and astronomical data furnished by superior conjunctions of Venus (1966) and Mercury (1967). Initially, the effect was verified with the precision of about 20%. The experiments have been repeated many times since then, with increasing accuracy, up to 1%.

Chapter 7

Lagrangian Approach in Invariantive Mechanics

7.1. Preliminaries

In view of the equivalence between Lagrangian and canonical formulations (see Chap. 1), we now propose ourselves to apply the Lagrangian method in invariantive mechanics. This formalism is imposed by the fact that the equations of motion deduced in Chapter 5 cannot be applied for $v \equiv c$, while the Lagrangian procedure avoids the series expansion with respect to v/c and allows to consider the asymptotic case $v \to c$.

The Lagrangian method in invariantive mechanics was first used by N.I.Pallas [30] and L.Sofonea [31]. According to these authors, the Lagrangian depends only on the Euclidean invariants $\mathbf{v}_1^2, \mathbf{v}_2^2, \mathbf{v}_1 \cdot \mathbf{v}_2, \mathbf{v}_1 \cdot \mathbf{r}, \mathbf{v}_2 \cdot \mathbf{r}$ and r, on the conditions

$$\frac{\partial L_\infty}{\partial (\mathbf{v}_1 \cdot \mathbf{v}_2)} = 0;$$

$$\frac{\partial^2 L_\infty}{\partial (v_1^2) \partial (v_2^2)} = 0,$$

where L_∞ stands for the Lagrangian with $r \to \infty$. The Lagrangian then becomes

$$L = \mu_1 g_1(\mathbf{v}_1^2) + \mu_2 g_2(\mathbf{v}_2^2)$$

$$+ q_1 q_2 f(\mathbf{v}_1^2, \mathbf{v}_2^2, \mathbf{v}_1 \cdot \mathbf{v}_2, \mathbf{v}_1 \cdot \mathbf{r}, \mathbf{v}_2 \cdot \mathbf{r}, |\mathbf{r}|),$$

99

where parameters μ_1, μ_2 characterize the inertial properties of the bodies, q_1, q_2 are coupling parameters describing the interaction with other bodies, while f, g_1, g_2 are some adjustment functions.

Using Euler-Lagrange equations of motion, the above mentioned authors obtained certain expressions for the centre-of-mass radius vector, the energy and the linear momentum of the system. The same method is applied to study the motion of a two-body system in the fields, but the obtained Lagrangian functions are not used to give a concrete description of the motion (trajectory, etc.).

The procedure developed in this chapter is different from that offered by [30], [31], since our Lagrangian is composed in a direct manner, by means of momenta and energy formulas deduced in Chapter 5. In addition, the Lagrangian method shall be effectively applied to study the motion within the field formalism [6], [7], [27], [28].

7.2. Construction of the Lagrangian

The Lagrangian of the system, in invariantive mechanics, is obtained as (we assume, again, that the dot product sign is omitted)

$$L = \sum_{l=1}^{2} \mathbf{p}_l \mathbf{v}_l - H, \tag{7.1}$$

with:

$$\mathbf{p}_1 = m_1 \mathbf{v}_1 + \mu \mathbf{v}_2 + h_1 \kappa \mathbf{r}; \quad \mathbf{p}_2 = m_2 \mathbf{v}_2 + \mu \mathbf{v}_1 + h_2 \kappa \mathbf{r};$$

$$H = (m_1 + m_2 + 2\mu + 2\kappa)c^2 \ ;$$

$$m_1 = m_1^0 \left(1 - \frac{\mathbf{v}_1^2}{c^2}\right)^{-1/2} ; \quad m_2 = m_2^0 \left(1 - \frac{\mathbf{v}_2^2}{c^2}\right)^{-1/2} ;$$

$$\mu = -\frac{G m_1^0 m_2^0}{2c^2 r} \left(1 - \frac{\mathbf{v}_1 \mathbf{v}_2}{c^2}\right)^{-1/2} \left[1 + \frac{5}{2} G \frac{(m_1^0 + m_2^0)}{c^2 r}\right];$$

$$\kappa = -\frac{1}{2} g \ell_{12} \left(\frac{r}{c^2}\right)^2 (1 - u)^{-1/2};$$

$$u = \frac{m_1^0 v_1^2 \cos^2(\mathbf{v}_1, \mathbf{r}) + m_2^0 v_2^2 \cos^2(\mathbf{v}_2, \mathbf{r})}{2(m_1^0 + m_2^0)c^2}.$$

(7.2a–h)

The Lagrangian therefore is

$$L = -m_1^0 c^2 \left(1 - \frac{v_1^2}{c^2}\right)^{1/2} - m_2^0 c^2 \left(1 - \frac{v_2^2}{c^2}\right)^{1/2} - (2\mu + 2\kappa)c^2.$$

(7.3)

In order to study the motion of the body of mass m_2^0 relative to the body of mass m_1^0 let us consider a reference frame with invariable orientation and origin at body of mass m_1^0. Supposing that $m_2^0 \ll m_1^0$, the chosen frame is a good approximation to an inertial one. In this case, $\mathbf{v}_1 = 0$. Let us denote

$$m_1^0 = M_0, \quad m_2^0 = m_0, \quad \mathbf{v}_2 = \mathbf{v}$$

(7.4)

and call, for simplicity, the body of mass m_0 — probe-body, and that of mass M_0 — source-body.

The Lagrangian (7.3), by omitting the additive constant $M_0 c^2$, therefore is:

$$L = -m_0 c^2 \left(1 - \frac{\mathbf{v}^2}{c^2}\right)^{1/2} + \frac{GM_0 m_0}{r} \left(1 + \frac{5}{2}\frac{GM_0}{c^2 r}\right)$$
$$+ g\ell_{12}\frac{r^2}{c^2}\left[1 - \frac{m_0 v^2 \cos^2(\mathbf{v}, \mathbf{r})}{2M_0 c^2}\right].$$

(7.5)

7.3. Infragalactic metrics

For $\kappa \to 0$, that is neglecting the dilatational mass with respect to the gravitational mass μ, the Lagrangian (7.5) becomes

$$L \approx -m_0 c^2 (1 - \beta^2)^{1/2} + \frac{n}{r}\left(1 + \frac{5n}{r}\right) m_0 c^2;$$

(7.6a)

$$\beta^2 = \frac{\mathbf{v}^2}{c^2},$$

(7.6b)

If the energy equation is written in the form

$$\beta^2 \approx \frac{2n}{r}, \tag{7.7}$$

we still have:

$$L \approx -m_0 c^2 (1 - \beta^2)^{1/2} + \left(\frac{n}{r} + \frac{6n^2 - n^2}{2r^2} \right) m_0 c^2$$

$$\approx -m_0 c^2 (1 - \beta^2)^{1/2} + \frac{n}{r} \left(1 + \frac{3n}{r} \right) m_0 c^2 - \frac{n^2}{2r^2} m_0 c^2$$

$$\approx -m_0 c^2 (1 - \beta^2)^{1/2} + \left[\left(1 + \frac{3}{2}\beta^2 \right) \frac{n}{r} m_0 c^2 - \frac{n^2}{2r^2} m_0 c^2 \right]. \tag{7.8}$$

By means of relation

$$\frac{1}{c} \left(\frac{dS}{dt} \right) = -\frac{L}{m_0 c^2}, \tag{7.9}$$

one can define a metric invariant. To this end, using (7.8), we still have

$$\frac{1}{c} \left(\frac{dS}{dt} \right) \approx (1 - \beta^2)^{1/2} - \left[\left(1 + \frac{3}{2}\beta^2 \right) \frac{n}{r} - \frac{n^2}{2r^2} \right]. \tag{7.10}$$

Squaring this relation, one obtains

$$\frac{1}{c^2} \left(\frac{dS}{dt} \right)^2 \approx (1 - \beta^2) - 2 \left(1 - \frac{1}{2}\beta^2 \right) \left[\left(1 + \frac{3}{2}\beta^2 \right) \frac{n}{r} - \frac{n^2}{2r^2} \right] + \frac{n^2}{r^2}$$

$$\approx (1 - \beta^2) - 2(1 - \beta^2) \frac{n}{r} + \frac{2n^2}{r^2} \tag{7.11}$$

$$= \left(1 - \frac{2n}{r} + \frac{2n^2}{r^2} \right) - \beta^2 \left(1 + \frac{2n}{r} \right).$$

It then follows the invariant $(dS)^2$

$$dS^2 \approx \left(1 - \frac{2n}{r} + \frac{2n^2}{r^2} \right) (c\,dt)^2 - \left(1 + \frac{2n}{r} \right) (d\mathbf{r})^2. \tag{7.12}$$

Since estimation of the gravitational tests with metric (7.12) leads to results in agreement with astronomical observations [30], one can extend its validity over velocities $v \equiv c$. in which case it becomes

$$dS^2 = e^{-\frac{2n}{r}} c^2 dt^2 - e^{\frac{2n}{r}} (dr^2 + r^2 d\theta^2 + r^2 \sin^2 \theta d\varphi^2). \qquad (7.13)$$

As one observes, expansion of exponential functions in (7.13) and retention of terms of the first power in n/r, lead to (7.12). Such an extension of a metric valid for $v \ll c$ to $v \equiv c$ is not singular: it was also used in general relativity to construct a weakly-covariant Lorentz metric (see [30] for details).

7.4. Differential equation of infragalactic motion

Suppose that the Sun, of mass M_0, generates a gravitational field described by the metric (7.13). A planet gravitating towards the Sun describes a geodesic of the Riemannian space. Let us determine the trajectory of such a planet.

To solve the problem, we should solve the system of differential equations obtained by applying the principle

$$\delta \int F \, d\tau = 0,$$

with

$$F = e^{-\frac{2n}{r}} c^2 - e^{\frac{2n}{r}} (\dot{r}^2 + r^2 \dot{\theta}^2 + r^2 \sin^2 \theta \dot{\varphi}^2) \equiv c^2. \qquad (7.14)$$

Since this procedure is difficult, we shall admit some simplifying hypotheses, based on astronomical observations:

i) Under the action of a central-type force, a planet describes a plane trajectory ($\varphi = const.$, $\dot{\varphi} = 0$);
ii) The planet trajectory is, in the first approximation, an ellipse with the Sun at one of the foci.

This approximation allows one to calculate the deviation, without rigorously solving Euler-Lagrange equations.

Since the independent variables are r, θ and t, we need three differential equation in these variables. The equations in t and θ are obtained by applying Euler-Lagrange equations to function

$$F = e^{-\frac{2n}{r}} c^2 \dot{t}^2 - e^{\frac{2n}{r}} \left(\dot{r}^2 + r^2 \dot{\theta}^2 \right) \equiv c^2. \tag{7.15}$$

We then have:

$$\frac{\partial F}{\partial \dot{t}} = 2c^2 e^{-\frac{2n}{r}} \dot{t} = 2\varepsilon_0; \tag{7.16a}$$

$$\frac{\partial F}{\partial \dot{\theta}} = -2c^2 \dot{\theta} e^{\frac{2n}{r}} = -2L_0, \tag{7.16b}$$

where ε_0 and L_0 are constants. As equation for r can be considered (7.15). The system of equations used to determine r, θ, t in terms of τ, with $dS = cd\tau$, then is

$$e^{-\frac{2n}{r}} c^2 \dot{t}^2 - e^{\frac{2n}{r}} \left(\dot{r}^2 + r^2 \dot{\theta}^2 \right) \equiv c^2; \tag{7.17a}$$

$$e^{-\frac{2n}{r}} \ddot{t} = \varepsilon_0; \tag{7.17b}$$

$$r^2 \dot{\theta} e^{\frac{2n}{r}} = L_0. \tag{7.17c}$$

Let us set up the differential equation of trajectory. Observing that

$$\dot{r} = \frac{dr}{d\tau} = \frac{dr}{d\theta} \frac{d\theta}{d\tau} = \frac{L_0}{r^2} e^{-\frac{2n}{r}} \frac{dr}{d\theta} = -L_0 e^{-\frac{2n}{r}} \frac{d}{d\theta} \left(\frac{1}{r} \right) \tag{7.18}$$

and eliminating $\dot{t}$ and $\dot{\theta}$ between (7.17a–c) and using (7.18), we still have

$$\left(\frac{du}{d\theta} \right)^2 + u^2 = - \left(\frac{c}{L_0} \right)^2 e^{2nu} + \left(\frac{c\varepsilon_0}{L_0} \right)^2 e^{4nu}; \quad u = \frac{1}{r}, \tag{7.19}$$

or, if derivative with respect to u is performed,

$$u'' + u = -n \left(\frac{c}{L_0} \right)^2 e^{2nu} + 2n \left(\frac{c\varepsilon_0}{L_0} \right)^2 e^{4nu}. \tag{7.20}$$

7.5. Planetary perihelion advance

If $v \ll c$, we have $nu \ll 1$ and $\varepsilon_0 \approx 1$. In this case, equation (7.20) takes the approximate form

$$u'' + u\left[1 - \left(\frac{6nc}{L_0}\right)^2\right] \approx n\left(\frac{c}{L_0}\right)^2. \tag{7.21}$$

Next, using procedure described in Chapter 5, one finds the result given by astronomical observations.

7.6. Deflection of light in gravitational field

Consider a light beam which propagates in the Sun's gravitational field. Its path, according to general relativity, is a geodesic of the Riemannian space, characterized by $dS = 0$ (the so-called null geodesic). Supposing that the trajectory lies in a plane and using (7.13), we have:

$$e^{-\frac{2n}{r}}c^2dt^2 - e^{\frac{2n}{r}}(dr^2 + r^2d\theta^2) = 0; \tag{7.23}$$

$$e^{\frac{4n}{r}}r^2d\theta = \frac{L_0}{\varepsilon_0}dt. \tag{7.24}$$

Eliminating dt between the last two equations, one obtains

$$\left(\frac{du}{d\theta}\right)^2 + u^2 = \left(\frac{c\varepsilon_0}{L_o}\right)^2 e^{4nu}, \tag{7.24}$$

or, taking the derivative

$$u'' + u = 2n\left(\frac{c\varepsilon_0}{L_o}\right)^2 e^{4nu}. \tag{7.25}$$

For $nu \ll 1$ and $\dfrac{c\varepsilon_0}{L_0} \to \dfrac{1}{R^2}$, equation (7.25) becomes

$$u'' + u = \frac{2n}{R^2}. \tag{7.26}$$

For further calculation, one applies method furnished by Chapter 6. The result is in concordance with the astronomical observations.

7.7. The red shift of spectral lines in strong gravitational fields

Consider a punctiform mass M_0 whose gravitational field is described by the metric (7.13). A horologe connected to this system ($dr = 0$, $d\theta = 0$, $d\varphi = 0$) determines the proper duration

$$\Delta\tau = \int \frac{dS}{c} = \int_{t_1}^{t_2} e^{-\frac{2n}{r}} \, dt = e^{-\frac{2n}{r}} \Delta t. \qquad (7.27)$$

In another point of the same frame ($dr' = 0$, $d\theta = 0$, $d\varphi = 0$) an identical horologe, during the same duration Δt, registers the proper duration

$$\Delta\tau' = e^{-\frac{2n}{r'}} \Delta t. \qquad (7.28)$$

The last two relations give

$$e^{\frac{2n}{r}} \Delta\tau = e^{\frac{2n}{r'}} \Delta\tau'. \qquad (7.29)$$

Suppose that the first horologe lies on the surface of a star of radius R and the second on the surface of the Earth, situated at the distance $r' \gg R$. In this case one can approximate

$$\Delta\tau' \approx e^{\frac{2n}{R}} \Delta\tau. \qquad (7.30)$$

Such identical horologes exist in Universe as atoms of any element, which, according to astronomical observations, are identical. Let ν be the frequency of the radiation emitted by an atom situated on the considered star. In agreement with (7.30), the radiation frequency determined from the Earth is

$$\nu' = e^{-\frac{2n}{r}} \nu \approx \left(1 - \frac{n}{R}\right) \nu < \nu', \qquad (7.31)$$

corresponding to a red displacement of the spectral line associated with the frequency ν emitted by the atom.

7.8. Delay of laser echo

Condition $dS = 0$ implies, via (7.13), the dependence of velocity of light waves upon intensity of the gravitational field traveled by the light

$$\beta^2 = \frac{v^2}{c^2} = e^{-\frac{4n}{r}}. \tag{7.32}$$

For $\frac{n}{r} \ll 1$, and taking into account only the one-dimensional case, one obtains the index of refraction

$$n = \frac{c}{v} \approx 1 + \frac{2n}{r}. \tag{7.33}$$

Therefore, a static gravitational field behaves like a medium with refractive index given by (7.33). From now on, following the same procedure as in Chapter 6, one can investigate the gravitational delay of the light beam.

Chapter 8

Considerations on Invariantive Mechanics

8.1. Preliminaries

The first aspect of our analysis concerns the principles of invariantive mechanics. One of these principles identifies with that of classical mechanics connected to the structure $E_3 \times T$ of the space-time variety, while the other one enlarges the frame of special relativity by admitting some limited propagation velocities, which could differ from the speed of light.

The motion principle expressed by closeness of certain 1-forms is more general than the differential principles of classical mechanics and it is already widely used in mechanics [3], [5], [15].

Another aspect refers to functional dependencies of the Hamiltonian on Euclidean invariants. As long as the inertial transformations emerge from a certain model they can be accepted. On the contrary, if the inertial transformations are postulated, one has to give up these dependencies, because any Lorentz invariant analytical function of a vector system, as the Hamiltonian can be, is an analytic function of the invariants which can be constructed with the considered vector system [15]. But the last solution implies some already known physical models: special relativity, gauge theories, etc.

It then remains to analyze the possibility of considering the Hamiltonian as being dependent of a smaller number of Euclidean invariants, to simplify the mathematical treatment of the problem. Indeed, two material points form a system only if they reciprocally interact. This fact is valid only for attractive interactions

109

(Newtonian, invariantive, Coulombian, etc.). In case of repulsive forces, such as dilatational interaction introduced by invariantive mechanics, repulsive interaction increases with distance and tends to infinity. In this context, motivation for existence of the Repulsive force by means of Hubble's law doesn't seem to be sufficient. On the other hand, it is difficult to accept conclusion of the author of invariantive mechanics, Octav Onicescu, according to which the Universe has to be finite, in order to avoid the unlimited increase of the repulsive force. It is also true that the Newtonian force is subject to the singularity $r \to 0$. Besides, if the Universe would be finite, then would appear the gravity paradox. In conclusion, the problem exceeds the frame of invariantive mechanics and it is far from being solved.

According to invariantive mechanics, any two bodies (considered as material points) form a system: attraction is preponderant at small distances, and repulsion at big distances. Since those two forces cannot unlimitedly increase, the distance between the two points must have an upper and lower limits. If the Universe is not considered finite and the invariantive mechanics principles are accepted, then parameter ℓ_{12} associated to the system has to compensate the effect of increasing of distance. The simplest solution is to consider the dilatational mass as being zero.

The problem is to decide on which Euclidean invariants the energy H depends. In principle, H can depend on all the Euclidean invariants, but one must consider only those invariants which are significant for the system. Such invariants were used by Weber and Helmholtz [31].

The multitude of two material points form a system only if the distance between them is finite. On the other hand, displacement of the system is given by a certain asymptotic peculiarity of the Hamiltonian, namely

$$H = H(\alpha_1, \alpha_2), \tag{8.1}$$

meaning that the distance between them is infinite. This way, by dissociation of the system, the mass of gravitational interaction μ as well as dilatational mass κ receive a latent existence, not being present in the limited functions $\mathbf{p}_1, \mathbf{p}_2, H$:

$$\mathbf{p}_1 = m_1\mathbf{v}_1; \quad \mathbf{p}_2 = m_2\mathbf{v}_2; \quad H = (m_1 + m_2)c^2. \qquad (8.2)$$

Dependence of the Hamiltonian H on parameters μ and κ, which represent a measure of coupling of the system, must be settled so as to express the extensive character of these parameters. This essential property of 'latent variables' can be represented by functional relation

$$H = H(\alpha_1, \alpha_2, \alpha, \beta_1, \beta_2, \beta). \qquad (8.3)$$

Even if relation (8.3) implies an incomplete determination of parameters μ and κ — this situation being specific to any phenomenological mass — it does not exclude the interdependence of the two types of interactions. But this reasoning contradicts the hypothesis according to which the two types of interactions are distinct forms on inertia. On the other side, even identification of interactions compatible with the inertial structure of the system of two material points proves to be subjective: due to arbitrarity of functions $\varphi(r)$ and $\psi(r)$, the definition of gravitational mass as a dilatational mass and vice-versa is equally justified as well. Taking into account these circumstances, we have tried to eliminate such a situation from the system of invariantive mechanics, using one of the following selection rules for the Euclidean invariants:

i)

$$H = H(\alpha_1, \alpha_2, \alpha, \beta), \qquad (8.4)$$

in which case

$$\mathbf{p}_1 = m_1\mathbf{v}_1 + \mu\mathbf{v}_2; \quad \mathbf{p}_2 = m_2\mathbf{v}_2 + \mu\mathbf{v}_1;$$

$$H = (m_1 + m_2 + 2\mu)c^2;$$

$$\frac{d\mathbf{p}_1}{dt} = -\mathbf{L}; \quad \frac{d\mathbf{p}_2}{dt} = \mathbf{L}; \qquad (8.5)$$

$$\mathbf{L} = 2c^2 \left(1 - \frac{\mathbf{v}_1 \cdot \mathbf{v}_2}{c^2} \right) \frac{\partial \mu}{\partial r} \frac{\mathbf{r}}{r};$$

$$\mu = \frac{\varphi(r)}{\left(1 - \frac{\mathbf{v}_1 \cdot \mathbf{v}_2}{c^2} \right)^{1/2}}.$$

ii)

$$H = H(\alpha_1, \alpha_2, \beta_1, \beta_2), \tag{8.6}$$

followed by

$$\mathbf{p}_1 = m_1 \mathbf{v}_1 + h_1 \kappa \mathbf{v}_2; \quad \mathbf{p}_2 = m_2 \mathbf{v}_2 + h_2 \kappa \mathbf{v}_1;$$

$$H = (m_1 + m_2 + 2\kappa)c^2;$$

$$\frac{d\mathbf{p}_1}{dt} = -\mathbf{L}; \quad \frac{d\mathbf{p}_2}{dt} = \mathbf{L}; \tag{8.7}$$

$$\mathbf{L} = \left(2c^2 \frac{\partial \kappa}{\partial r} - \mathbf{v}_1 \mathbf{r} \frac{\partial \ell_1}{\partial r} - \mathbf{v}_2 \mathbf{r} \frac{\partial \ell_2}{\partial r} \right) \frac{\mathbf{r}}{r} - \ell_1 \mathbf{v}_1 - \ell_2 \mathbf{v}_2;$$

$$\kappa = \frac{\psi(r)}{(1-u)^{1/2}}; \quad u = \frac{m_1^0 v_1^0 \cos^2(\mathbf{v}_1, \mathbf{r}) + m_2^0 v_2^0 \cos^2(\mathbf{v}_2, \mathbf{r})}{2(m_1^- + m_2^0)c^2}.$$

But this last selection rule cannot (even in principle) be discussed, because the implied functional equation can only formally be solved. Consequently, the physical picture of invariantive mechanics must be reinterpreted on the basis of selection rule i), with the mention of arbitrarity of the mass function; it can be suppressed by imposing some supplementary hypotheses on the mass function $\varphi(r)$, so as to allow a complete determination of the interaction masses as well as a veridic explanation of gravitational effects at infragalactic scale (planetary perihelion advance, light deviation in gravitational field, red shift in intense gravitational fields, delay of radar signals, etc.) and galactic dimensions (Hubble effect).

8.2. Mass of gravitational interaction

According to the investigation developed in Chapter 5, let us choose $\varphi(r)$ as

$$\varphi(r) = \left\{ -\frac{1}{2}\frac{Gm_1^0 m_2^0}{c^2 r}\left[1 + \frac{5}{2}\frac{G(m_1^0 + m_2^0)}{c^2 r}\right] \right.$$
$$\left. +\frac{1}{4}\frac{m_1^0 m_2^0}{m_1^0 + m_2^0}\left(\frac{r}{cT}\right)^2 \right\}, \tag{8.8}$$

in which case the mass of gravitational interaction writes

$$\mu(r) = \left\{ -\frac{1}{2}\frac{Gm_1^0 m_2^0}{c^2 r}\left[1 + \frac{5}{2}\frac{G(m_1^0 + m_2^0)}{c^2 r}\right] \right.$$
$$\left. +\frac{1}{4}\frac{m_1^0 m_2^0}{m_1^0 + m_2^0}\left(\frac{r}{cT}\right)^2 \right\}\left(1 - \frac{\mathbf{v}_1\mathbf{v}_2}{c^2}\right)^{-1/2}. \tag{8.9}$$

Therefore, interaction between any two bodies of the Universe is manifested as:

a) infragalactic gravitational mass

$$\mu_i(r) = -\frac{1}{2}\frac{Gm_1^0 m_2^0}{c^2 r}\left[1 + \frac{5}{2}\frac{G(m_1^0 + m_2^0)}{c^2 r}\right]; \tag{8.10}$$

b) extragalactic gravitational mass

$$\mu_l(r) = \frac{1}{4}\frac{m_1^0 m_2^0}{m_1^0 + m_2^0}\left(\frac{r}{cT}\right)^2, \tag{8.11}$$

where the epoch T and Hubble's constant H are connected by the relation

$$T = H^{-1}.$$

These two species of mass are simultaneously present, with observation that the attractive one (8.10) is dominant at infragalactic scale, such as the solar system, while the repulsive mass acts on the extragalactic scale.

8.3. Hubble's effect

Under this alternative, with approximation $|\mathbf{v}_l| \ll c$, the equations of relative motion

$$m_1^0 \left(\frac{d^2\mathbf{r}_1}{dt^2} - \frac{\mathbf{r}_1}{T^2} \right) = -\frac{Gm_1^0 m_2^0}{r^3}\mathbf{r};$$

$$m_2^0 \left(\frac{d^2\mathbf{r}_2}{dt^2} - \frac{\mathbf{r}_2}{T^2} \right) = \frac{Gm_1^0 m_2^0}{r^3}\mathbf{r} \tag{8.12}$$

lead to the equations of relative motion

$$\ddot{\mathbf{r}} - \frac{1}{T^2}\mathbf{r} = -G(m_1^0 + m_2^0)\frac{\mathbf{r}}{r^3}. \tag{8.13}$$

Relation (8.13) is invariant with respect to the group of transformations [30], [32]

$$\mathbf{r}' = \mathbf{r} + \mathbf{v}_0 T \sinh\left(\frac{t - t_0}{T}\right); \quad t' = t, \tag{8.14}$$

with

$$\mathbf{v}_0; \quad t_0 = const. \tag{8.15}$$

One can distinguish two limit cases:

a) $T \to \infty$, also called infragalactic case, in which situation (8.13) reduces to the Newtonian equation

$$\ddot{\mathbf{r}} = -G(m_1^0 + m_2^0)\frac{\mathbf{r}}{r^3}$$

and (8.14) to Galilean group

$$\mathbf{r}' = \mathbf{r} + \mathbf{v}_0 t; \quad t' = t. \tag{8.16}$$

b) $G \to 0$, known as extragalactic case, when (8.13) goes to

$$\ddot{\mathbf{r}} = \frac{1}{T^2}\mathbf{r}, \tag{8.17}$$

with solution

$$\mathbf{r} = \mathbf{r}_0 \cosh\left(\frac{t - t_0}{T}\right) + \mathbf{v}_0 T \sinh\left(\frac{t - t_0}{T}\right), \tag{8.18}$$

with

$$\mathbf{r}_0; \quad \mathbf{v}_0 = const.$$

If duration of the motion is much smaller then the constant T, then $(t - t_0)/T \ll 1$ and the last equation describes inertial motion in Galilean formulation

$$\mathbf{r} = \mathbf{r}_0 + \mathbf{v}_0(t - t_0).$$

For $(t - t_0)/T \gg 1$, equation (8.18) has a limit. This limit can be found by taking the time derivative of (8.18) and then eliminating $\mathbf{r}_0$ between (8.18) and the obtained derivative. The result is

$$\mathbf{v} = \mathbf{v}_0 \cosh^{-1}\left(\frac{t - t_0}{T}\right) + \mathbf{r}T^{-1} \tanh\left(\frac{t - t_0}{T}\right). \tag{8.19}$$

Taking now the limit $t - t_0 \gg T$, we are left with Hubble's law

$$\mathbf{v} = H\,\mathbf{r}. \tag{8.20}$$

8.4. Invariantive model with epoch-dependent G

Let us now perform the space-time transformation

$$\tau = T \tanh\left(\frac{t - t_0}{T}\right);$$

$$\boldsymbol{\rho} = \frac{\mathbf{r}}{\cosh\left(\frac{t - t_0}{T}\right)}. \tag{8.21}$$

Under this transformation, (8.14) goes to the Galilean group

$$\boldsymbol{\rho}' = \boldsymbol{\rho} + \mathbf{v}_0 \tau; \quad \tau' = \tau. \tag{8.22}$$

We also have the identity:

$$\frac{d^2\boldsymbol{\rho}}{d\tau^2} \equiv \left(\frac{d^2\mathbf{r}}{dt^2} - \frac{\mathbf{r}}{T^2}\right)\cosh^3\left(\frac{t-t_0}{T}\right). \tag{8.23}$$

It follows from (8.17) and (8.23) that in the measure $(\boldsymbol{\rho},\tau)$ a body acted only by the matter (of smoothed distribution) of the Universe, moves rectilinearly and uniformly. In other words, the effect of matter of the Universe upon the probe body (free of local actions) are compensated in an exact manner, meaning that the matter motion has only ergodic component. Consequently, under transition from measure $(\mathbf{r},t)$ to measure $(\boldsymbol{\rho},\tau)$, the ordered (hydrodynamic) component of motion vanish. This way, transition $(\mathbf{r},t) \to (\boldsymbol{\rho},\tau)$ is equivalent to transition from the observation frame to the co-mobile frame specific to conventional cosmology [30].

Let us now insert in (8.12a,b) the transformations

$$t = t_0 + T\ln\left(\frac{T+\tau}{T-\tau}\right)^{1/2};$$

$$\mathbf{r} = \frac{\boldsymbol{\rho}}{\left(1 - \frac{\tau^2}{T^2}\right)^{1/2}}, \tag{8.24}$$

which is the inverse of transformation (8.12a,b), One obtains:

$$m_1^0 \frac{d^2\boldsymbol{\rho}_1}{d\tau^2} = \frac{G}{\left(1 - \frac{\tau^2}{T^2}\right)^{1/2}} \frac{m_1^0 m_2^0}{\rho^3}\boldsymbol{\rho};$$

$$m_2^0 \frac{d^2\boldsymbol{\rho}_2}{d\tau^2} = -\frac{G}{\left(1 - \frac{\tau^2}{T^2}\right)^{1/2}} \frac{m_1^0 m_2^0}{\rho^3}\boldsymbol{\rho}. \tag{8.25}$$

Since in the measure $(\boldsymbol{\rho},\tau)$ is valid the Galilei's inertia principle, equations (8.25a,b) have to coincide with Newton's equations. Consequently, the constant G appearing in (8.25) must depend on epoch according to the law

$$G = G_0\left(1 - \frac{\tau^2}{T^2}\right)^{1/2} = \frac{G_0}{\cosh\left(\frac{t-t_0}{T}\right)}. \tag{8.26}$$

Even if relation (8.26) shows a decrease of G with respect to epoch, this result is considerably different from Dirac and Dicke cosmology (see [30] for details), because $(\dot{G}/G) = 0$ for $t = t_0$.

8.5. Hubble's effect in the high-speed approximation

Consider the linear momentum conservation law

$$\mathbf{p}_1 + \mathbf{p}_2 = const. \tag{8.27}$$

or, explicitly

$$m_1\mathbf{v}_1 + \mu\mathbf{v}_2 + m_2\mathbf{v}_2 + \mu\mathbf{v}_1 = \mathbf{c}. \tag{8.28}$$

Multiplying by $\mathbf{r}/r$ and recalling that

$$\mathbf{v}_1\mathbf{r} = r\xi_1; \quad \mathbf{v}_2\mathbf{r} = r\xi_2;$$

$$\xi_1 = v_1\cos(\mathbf{v}_1,\mathbf{r}); \quad \xi_2 = v_2\cos(\mathbf{v}_2,\mathbf{r}), \tag{8.29}$$

we still have

$$m_1\xi_1 + m_2\xi_2 + \mu(\xi_1 + \xi_2) = \cos(\mathbf{c},\mathbf{r}). \tag{8.30}$$

Using the expression for μ and denoting

$$M = -\frac{1}{2}\frac{Gm_1^0 m_2^0}{\sqrt{1 - \frac{\mathbf{v}_1\mathbf{v}_2}{c^2}}}\frac{\xi_1 + \xi_2}{m_1\xi_1 + m_2\xi_2 - c\cos(\mathbf{c},\mathbf{r})}$$

$$\times \left[1 + \frac{5}{2}\frac{G(m_1^0 + m_2^0)}{c^2 r}\right]; \tag{8.31}$$

$$N = \frac{1}{4}\frac{m_1^0 m_2^0}{m_1^0 + m_2^0}\frac{1}{\sqrt{1 - \frac{\mathbf{v}_1\mathbf{v}_2}{c^2}}}\frac{\xi_1 + \xi_2}{m_1\xi_1 + m_2\xi_2 - c\cos(\mathbf{c},\mathbf{r})} \tag{8.32}$$

we have

$$1 + \frac{M}{c^2 r} + \frac{1}{c^2 T^2}Nr^2 = 0. \tag{8.33}$$

Taking the time derivative, we obtain:

$$-\frac{2M}{c^2 r^4}\mathbf{r}\dot{\mathbf{r}} + \frac{\dot{M}}{c^2 r} + \frac{1}{c^2 T^2}\dot{N}r^2 + \frac{1}{c^2 T^2}2N\mathbf{r}\dot{\mathbf{r}}\,. \tag{8.34}$$

Multiplying by $\frac{c^2 T^2}{2Nr}$, we still have

$$-\frac{MT^2}{NR^5}\mathbf{r}\dot{\mathbf{r}} + \frac{\dot{M}T^2}{2Nr^3} + \frac{1}{2}\frac{\dot{N}}{N}r + \frac{1}{r}\mathbf{r}\dot{\mathbf{r}} = 0, \tag{8.35}$$

or, after some term arrangements,

$$\frac{\mathbf{r}\dot{\mathbf{r}}}{r}\left(1 - \frac{MT^2}{NR^4}\right) = -\frac{1}{2}\frac{\dot{N}}{N}r\left(1 + \frac{\dot{M}}{\dot{N}}\frac{T^2}{r}\right). \tag{8.36}$$

The final result is

$$|\mathbf{v}_2\cos(\mathbf{v}_2,\mathbf{r}) - \mathbf{v}_1\cos(\mathbf{v}_1,\mathbf{r})| = \frac{1}{2}\left|\frac{\dot{N}}{N}\right|r\left|\frac{1 + \frac{\dot{M}}{\dot{N}}\frac{T^2}{r}}{1 - \frac{MT^2}{Nr^4}}\right|, \tag{8.37}$$

or, still

$$|\mathbf{v}_2\cos(\mathbf{v}_2,\mathbf{r}) - \mathbf{v}_1\cos(\mathbf{v}_1,\mathbf{r})| = \theta(r)\,r \tag{8.38}$$

where

$$\theta(r) = \frac{1}{2}\left|\frac{\dot{N}}{N}\right|\left|\frac{1 + \frac{\dot{M}}{\dot{N}}\frac{T^2}{r}}{1 - \frac{MT^2}{Nr^4}}\right|. \tag{8.39}$$

The fact that θ is a function of r, which means the dependence $T = T(r)$, shows a certain ranking (fractality) of matter in the Universe.

8.6. Invariantive models of the universe

Consider a spherical universe of radius $r = R$ and density $\rho_0 = const.$ Then

$$m_0 = m_2^0 + m_2^0 = \frac{4\pi}{3}\rho_0 R^3 \tag{8.40}$$

and (8.13) becomes

$$\ddot{\mathbf{R}} + \frac{4\pi\Gamma\rho_0}{3}\left(1 - \frac{3H^2}{4\pi G\rho_0}\right)\mathbf{R} = 0. \qquad (8.41)$$

Suppose

$$\rho_0 > \rho_c = \frac{3H^2}{4\pi G}. \qquad (8.42)$$

In this case, solution of the equation (8.41), which has the form

$$\mathbf{R} = \mathbf{a}\sin(\omega_0 t + \rho) \qquad (8.43)$$

with

$$\mathbf{a} = const.; \quad \omega_0^2 = \frac{4\pi G\rho_0}{3}\left(1 - \frac{\rho_c}{\rho_0}\right); \quad \rho = const. \qquad (8.44)$$

defines the pulsatory model of the universe: the universe expansion attains a maximum, followed by contraction pointing to a new Big Bang, then evolution repeats cyclically, with period

$$T_0 = \frac{2\pi}{\omega_0}. \qquad (8.45)$$

If $\rho_0 = \rho_c$, then (8.41) describes the parabolic model of universe, while $\rho_0 < \rho_c$, the described model is the hyperbolic one. In both models, after the Big Bang explosion, the universe expands to infinity, but the expansion rate is higher in the hyperbolic model.

8.7. Generalized Lorentz transformations. Critical mass of the universe

Insertion of relativistic amendments has to be done in such a way that equation (8.17), prescribing the extended inertial character of motion of a body (free of local actions), remains unchanged. Here is the solution:

a) Transformations (8.22a,b) have to be replaced by Lorentz transformations in measure $(\boldsymbol{\rho}, \tau)$;

b) Rewriting these transformations in measure $(\mathbf{r}, t)$.

It then follows the group of transformations

$$\mathbf{r} = (1 - \varepsilon^2)^{-1/2} \left[\frac{\mathbf{r}}{\cosh \alpha} + \gamma \mathbf{v}_0 T \tanh \alpha + \frac{\gamma^2}{\gamma + 1} \frac{(\mathbf{r} \cdot \mathbf{v}_0)\mathbf{v}_0}{c^2 \cosh \alpha} \right];$$

$$\tag{8.46}$$

$$t' = t_0 + T \ln \left(\frac{1 + \varepsilon}{1 - \varepsilon} \right)^{1/2},$$

with

$$\gamma = \left(1 - \frac{v_0^2}{\omega^2} \right)^{-1/2}; \quad \alpha = \frac{t - t_0}{T};$$

$$\tag{8.47}$$

$$\varepsilon \equiv \gamma \left(\tanh \alpha + \frac{1}{c^2} \frac{(\mathbf{r} \cdot \mathbf{v}_0)}{T \cosh \alpha} \right).$$

Transformations (8.46a,b), together with (8.47a,b), leave invariant the metric

$$dS^2 = \frac{dt^2}{\cosh^2 \alpha} \left[1 - \frac{1}{c^2} \left(\mathbf{v} \cosh \alpha - \frac{\mathbf{r}}{T} \sinh \alpha \right)^2 \right] \tag{8.48}$$

while the equations of motion (8.17) correspond to geodesic motion with metric (8.48). The universe dimensions are estimated that, in the measure $(\boldsymbol{\rho}, \tau)$, the matter aggregation takes place until the gravitational charge ρ_G of the universe becomes null:

$$\rho_G = \frac{\sqrt{G_0}}{c^2} \left(M_0 c^2 - \frac{3}{5} G_0 \frac{M_0^2}{cT} \right) = 0. \tag{8.49}$$

The inertia mass obtained this way is comparable to that estimated by A. Eddington and E. Milne [30].

8.8. Laws of variation of mass with time in invariantive mechanics

As is well known, Mescerski's laws regarding the variation of mass with the time were obtained in the hypothesis of existence of the attractive potential $V_a \sim 1/r$ [2]. The invariantive mechanics enlarges the classical frame by introducing the repulsive potential $V_r \sim r^2$, allowing generalization of these laws ([33], [34]).

Let us conceive the following simplifying hypotheses:

a) The total mass of the two-body system is an unknown function of time

$$m_1 + m_2 = m(t); \qquad (8.50)$$

b) The dominant interactions are those of repulsive and attractive type;

c) The displacement velocities $\mathbf{v}_\ell$ of the bodies satisfy the relation

$$|\mathbf{v}_\ell| \ll c. \qquad (8.51)$$

As a result, the equations of motion encountered in Chapter 5 become

$$m_1 \ddot{\mathbf{r}}_1 = G\frac{m_1 m_2}{r^3}\mathbf{r} - 2g\frac{m_1 m_2}{c^2}\mathbf{r}, \qquad (8.52)$$

and

$$m_2 \ddot{\mathbf{r}}_2 = - G\frac{m_1 m_2}{r^3}\mathbf{r} + 2g\frac{m_1 m_2}{c^2}\mathbf{r}, \qquad (8.53)$$

with

$$l_{12} - m_1 m_2. \qquad (8.54)$$

Multiplying (8.52) by m_1^{-1} and (8.53) by m_2^{-1}, then subtracting them side-by-side, one obtains the equation of relative motion:

$$\ddot{\mathbf{r}} + G\frac{m_1 + m_2}{r^3}\mathbf{r} = 2gc^{-2}(m_1 + m_2)\mathbf{r}. \qquad (8.55)$$

Since the trajectory lies in a plane, in Cartesian coordinate system (x_1, x_2), equation (8.55) writes

$$\frac{d^2 x_l}{dt^2} + \frac{Gmx_l}{r^3} = 2gc^{-2}mx_l; \quad (l = 1, 2), \qquad (8.56)$$

or, if the dimensionless parameters [2]

$$\xi_l^0 = \frac{x_l}{a}; \quad \tau_0 = \frac{t}{T}; \quad m' = \frac{m}{m_0}, \tag{8.57}$$

as well as the interaction constants

$$G_0 = \frac{m_0 G T^2}{a^3}; \quad K = 2gc^{-2}m_0 T^2 \tag{8.58}$$

are introduced, then one finds

$$\frac{d^2 \xi_l^0}{d\tau_0^2} + G_0 \frac{m' \xi_1^0}{(\xi_l^{02} + \xi_l^{02})^{3/2}} = K m' \xi_l^0. \tag{8.59}$$

To integrate (8.59), we shall follow the same steps as those shown in paper [33]. First, the set of variables ξ_l and τ are defined by

$$\xi_l^0 = m'^p \xi_l; \quad d\tau = m'^q d\tau_0, \tag{8.60}$$

where p and q are some arbitrary constants. Then we have

$$\frac{d\xi_l^0}{d\tau_0} = pm'^{p-1} \frac{dm'}{d\tau_0} \xi_l + m'^{p+q} \frac{d\xi}{d\tau}, \tag{8.61}$$

as well as

$$\frac{d^2 \xi_l^0}{d\tau_0^2} = p(p-1)m'^{p-2} \left(\frac{dm'}{d\tau_0}\right)^2 \xi_l + pm'^{p-1} \frac{d^2 m'}{d\tau_0^2} \xi_l$$

$$+ (2p+q)m'^{p+q-1} \left(\frac{dm'}{d\tau_0}\right) \left(\frac{d\xi_l}{d\tau}\right) + m'^{p+2q} \frac{d^2 \xi_l}{d\tau^2}. \tag{8.62}$$

Written in the new variables, equation (8.59) becomes

$$m'^{2q-2} + (2p+q)m'^{q-2} \left(\frac{dm'}{d\tau_0}\right) \left(\frac{d\xi_l}{d\tau}\right)$$

$$+ p(p-1)m'^{-3} \left(\frac{dm'}{d\tau_0}\right)^2 \xi_l$$

$$+ pm'^{-2} \frac{d^2 m'}{d\tau_0^2} \xi_l + G_0 m'^{-3p} \frac{\xi_l}{\rho^3} - K\xi_l = 0, \tag{8.63}$$

with

$$\rho^2 = \xi_1^2 + \xi_2^2, \tag{8.64}$$

To resume our task, we have to determine the unknown function m' for which (8.59), with transformation (8.60), admits integration of the system (8.63). To facilitate integration of (8.63), we choose p and q so that

$$2p + q = 0; \quad 2q - 1 = -3p, \tag{8.65}$$

which implies $p = 1$, $q = 2$. One finds:

$$m'^3 \left(\frac{d^2\xi_l}{d\tau^2} + G_0 \frac{\xi_l}{\rho^3} \right) = -\xi_l \left[\frac{2}{m'^3} \left(\frac{dm'}{\delta\tau_0} \right)^2 - \frac{1}{m'^2} \frac{d^2m'}{d\tau_0^2} - K \right]. \tag{8.66}$$

It follows that

$$\frac{2}{m'^3} \left(\frac{dm'}{\delta\tau_0} \right)^2 - \frac{1}{-m'^2} \frac{d^2m'}{\delta\tau_0^2} - K = c'm'^3, \tag{8.67}$$

where c' is a new constant. We then have

$$\frac{d^2\xi_l}{d\tau^2} + G_0 \frac{\xi_l}{\rho^3} + c'\xi_l = 0. \tag{8.68}$$

To integrate (8.67) one observes that

$$\frac{2}{m'^3} \left(\frac{dm'}{d\tau_0} \right)^2 - \frac{1}{m'^2} \frac{d^2m'}{d\tau_0^2} = m' \frac{d^2}{d\tau_0^2} \left(\frac{1}{m'} \right). \tag{8.69}$$

Therefore, solution can be written as

$$\tau_0 = \pm(2K)^{-1/2} \int \frac{z\,dz}{\sqrt{P(z)}}, \tag{8.70}$$

where

$$P(z) = z^3 + \frac{1}{2}\frac{c''}{K}z^2 - \frac{1}{2}\frac{c'}{K}, \tag{8.71}$$

where c'' is a new constant, and $z = m^{-1}$. We have arrived at an elliptic integral of the second kind (8.70) (for details see [35]). The sign is given by the sense of variation of variable z. This also means that the inverse of mass is an elliptic function of time.

For $K = 0$, which is equivalent to the absence of repulsion, and $c' = 0$, (8.67) reduces to Mescerski's law

$$m = (a\tau_0 + b)^{-1}, \tag{8.72}$$

when a and b are constants of integration. If $K = 0$ and $c' \neq 0$, we find

$$m = (a\tau_0 + ab + c'a^{-1})^{1/2}, \tag{8.73}$$

For $K \neq 0$ and $c' = 0$, by means of (8.57) one obtains the variation law

$$m = \frac{2K}{(K\tau_0 + a)^2 - c''}. \tag{8.74}$$

Chapter 9

Invariantive Mechanics of Rigid Body

9.1. General considerations

According to invariantive mechanics, a solid body is conceived as a discrete or continuous system of material points of finite space dimension and invariable distances between the points. The finite space extinction allows one to define the moments of inertia, whose calculation involves the velocity dependent mass distribution and fulfills the condition that the particle velocities are not much different from each other. This subject shall be developed by means of [5], [34].

9.2. Rigid body kinematics

To describe the motion of a rigid body in invariantive mechanics one usually employs two frames: one fixed, (X, Y, Z), and the other mobile, (x, y, z), invariably connected to the rigid. Most convenient is to choose the origin of the mobile frame at the centre of inertia of the body.

Position of the rigid body relative to the fixed frame is completely determined if the position of the mobile frame is known, Let $\mathbf{r}$ be the radius-vector of the origin O of the mobile frame. The orientation of axes of the mobile frame is determined by Euler's angles (θ, φ, ψ). This way, our system has six degrees of freedom, associated with the three components of $\mathbf{r}$ and Euler's angles (θ, φ, ψ).

An arbitrary infinitesimal displacement of the rigid body can be conceived as being composed by an infinitesimal displacement of

125

the centre of inertia (without changing the axes orientation), and an infinitesimal rotation of axes about the centre of inertia.

Let $\mathbf{r}$ be — this time — the position vector of some point P with respect to the mobile frame, and $\boldsymbol{\xi}$ the position vector of the same point relative to the fixed frame. According to the above considerations, an infinitesimal rotation is given by

$$d\mathbf{r} = d\boldsymbol{\alpha} \times \mathbf{r} \tag{9.1}$$

or

$$|d\mathbf{r}| = r \sin \beta \, d\alpha, \tag{9.2}$$

where β is the angle between $d\boldsymbol{\alpha}$ and $\mathbf{r}$. We then have

$$d\boldsymbol{\xi} = d\mathbf{R} + d\boldsymbol{\alpha} \times \mathbf{r}. \tag{9.3}$$

Multiplying by $(dt)^{-1}$ and defining

$$\mathbf{v} = \frac{d\boldsymbol{\xi}}{dt}; \quad \mathbf{V} = \frac{d\mathbf{R}}{dt}; \quad \boldsymbol{\Omega} = \frac{d\boldsymbol{\alpha}}{dt}, \tag{9.4}$$

we have

$$\mathbf{v} = \mathbf{V} + \boldsymbol{\Omega} \times \mathbf{r}. \tag{9.5}$$

Here $\mathbf{V}$ is the *velocity of the centre of inertia of the rigid*, and $\boldsymbol{\Omega}$ its *angular velocity*.

Suppose, now, that the origin of the mobile frame, invariably connected to the rigid body, is situate at a point O', different from the centre of inertia O, so that $\vec{OO'} = \mathbf{r_0}$. Let $\mathbf{V}$ be the velocity of O' and $\boldsymbol{\Omega}'$ the angular velocity. Consider some other point of the rigid, defined by $\mathbf{r}'$ with respect to O'. Since

$$\mathbf{r} = \mathbf{r}' + \mathbf{r_0}, \tag{9.6}$$

relation (9.5) becomes

$$\mathbf{v} = \mathbf{V} + \boldsymbol{\Omega} \times \mathbf{r_0} + \boldsymbol{\Omega} \times \mathbf{r}'. \tag{9.7}$$

On the other hand, according to (9.5)

$$\mathbf{v} = \mathbf{V}' + \boldsymbol{\Omega}' \times \mathbf{r}'. \tag{9.8}$$

It then follows that

$$\mathbf{V}' = \mathbf{V} + \boldsymbol{\Omega} \times \mathbf{r}_0; \quad \boldsymbol{\Omega} = \boldsymbol{\Omega}'. \tag{9.9}$$

Therefore, the angular velocity of the rigid body is independent of the coordinate system.

Under these circumstances, if $\mathbf{V}$ and $\boldsymbol{\Omega}$ are orthogonal in the frame with origin at O, $\mathbf{V}'$ and $\boldsymbol{\Omega}'$ shall be orthogonal with respect to some other origin O'. According to (9.5), velocities of the particles of the rigid are situated in the same plane, orthogonal to $\boldsymbol{\Omega}$. This means that one can always choose an origin O' so that $\mathbf{V}' = 0$, expressing a pure rotation about an axis passing through O', called *instantaneous axis of rotation*.

9.3. Inertia tensor of the rigid body

As well-known, the angular momentum of a body depends on the choice of the frame. If the origin of the fixed frame is chosen at the center of mass of the rigid body, then some specific properties and quantities can be evidentiated. According to our choice, $\mathbf{V} = 0$ and the body angular momentum identifies with the 'proper angular momentum', due to only the motion of the body with respect to its centre of inertia. Therefore, in the definition (here and hereafter, summation indices shall be omitted)

$$\mathbf{L} - \sum m(\mathbf{r} \times \mathbf{v}), \tag{9.10}$$

$\mathbf{v}$ must be replaced by

$$\mathbf{v} = \boldsymbol{\Omega} \times \mathbf{r} \tag{9.11}$$

which yields

$$\mathbf{L} = \sum m[\mathbf{r} \times (\boldsymbol{\Omega} \times \mathbf{r})]. \tag{9.12}$$

Since

$$\mathbf{r} \times (\boldsymbol{\Omega} \times \mathbf{r}) = r^2 \boldsymbol{\Omega} - \mathbf{r}(\boldsymbol{\Omega} \cdot \mathbf{r}), \tag{9.13}$$

we still have

$$\mathbf{L} = \sum m[r^2\mathbf{\Omega} - \mathbf{r}(\mathbf{\Omega} \cdot \mathbf{r})], \tag{9.14}$$

or, in tensor notation

$$L_i = \Omega_k \sum m[r_l^2\delta_{ik} - r_i r_k]; \quad (i, k, l = \overline{1,3}). \tag{9.15}$$

The quantities

$$I_{ik} = \sum m[r_l^2\delta_{ik} - r_i r_k] \tag{9.16}$$

form a symmetric tensor, called *inertia tensor*. If the mass distribution is continuous, the above sum is substituted by the integral

$$I_{ik} = \int (r_l^2\delta_{ik} - r_i r_k)dm. \tag{9.17}$$

If view of (9.15) and (9.16), we can write

$$L_i = I_{ik}\Omega_k. \tag{9.17'}$$

9.4. Equations of motion of the rigid body

Consider the 1-form associated with a solid (rigid) body, possessing a discrete mass distribution

$$\omega_s = \sum \mathbf{p}d\mathbf{r} + \sum \mathbf{L}d\boldsymbol{\alpha} - \left(\sum T + U\right) dt, \tag{9.18}$$

where $(d\mathbf{r}, d\boldsymbol{\alpha})$ define an infinitesimal roto-translation, and $\sum T + U$ is the Hamiltonian

$$H = \sum T + U, \tag{9.19}$$

with $\sum T$ — the kinetic energy, and U — the potential energy of the body situated in the field of external forces.

Postulate 9.1. *Motion of the rigid body is given by the closeness of the 1-form* (9.18), *that is*

$$D\omega_s = \sum d\mathbf{p} \wedge d\mathbf{r} + \sum d\mathbf{L} \wedge d\boldsymbol{\alpha} - \sum dT \wedge dt - \sum dU \wedge dt$$

$$= \sum d\mathbf{p}\delta\mathbf{r} - \sum \delta\mathbf{p}d\mathbf{r} + \sum d\mathbf{L}\delta\boldsymbol{\alpha} - \sum \delta\mathbf{L}d\boldsymbol{\alpha} \qquad (9.20)$$

$$- \sum dT\delta T + \sum \delta T dt - dU\delta t + \delta U dt = 0.$$

Since

$$\delta U = \sum \left(\frac{\partial U}{\partial \mathbf{r}}\delta\mathbf{r} + \frac{\partial U}{\partial \boldsymbol{\alpha}}\delta\boldsymbol{\alpha} \right), \qquad (9.21)$$

as a result of multiplication of (9.20) by $(dt)^{-1}$ and cancellation of coefficients of the arbitrary variations $\delta\mathbf{r}$, $\delta\boldsymbol{\alpha}.\,\delta t$, we are left with

a) the equations of motion of the rigid body

$$\sum \frac{d\mathbf{p}}{dt} = -\sum \frac{\partial U}{\partial \mathbf{r}}, \quad \sum \frac{d\mathbf{L}}{dt} = -\sum \frac{\partial U}{\partial \boldsymbol{\alpha}}; \qquad (9.22)$$

b) the energy conservation law

$$\frac{d}{dt}\left(\sum T + U \right) - 0, \quad \sum T + U - const.; \qquad (9.23)$$

c) the defining relation for the kinetic energy

$$\sum \delta T = \sum \boldsymbol{\Omega}\delta\mathbf{L} + \sum \mathbf{V}\delta\mathbf{p}, \qquad (9.24)$$

where presumption that all points of the rigid have the same $\boldsymbol{\Omega}$ and $\mathbf{V}$ has been used.

Relations (9.22) can be also written with respect to the centre of inertia. Consider an infinitesimal translation $\delta\mathbf{r}$ of the rigid body. Since the position vector $\mathbf{r}$ of each material point varies with the same quantity $\delta\mathbf{R}$, variation of the potential energy relative to $\mathbf{r}$ is

$$\delta_{\mathbf{r}}U = \sum \frac{\partial U}{\partial \mathbf{r}}\delta\mathbf{r} = \delta\mathbf{R}\sum \frac{\partial U}{\partial \mathbf{r}} = -\delta\mathbf{R}\sum \mathbf{f} = -\mathbf{F}\delta\mathbf{R}, \qquad (9.25)$$

where the force acting on a single particle $\mathbf{f}$, and the resultant force $\mathbf{F}$ are

$$\mathbf{f} = -\frac{\partial U}{\partial \mathbf{r}}; \quad \sum \mathbf{f} = \mathbf{F} = -\frac{\partial U}{\partial \mathbf{R}}. \tag{9.26}$$

Since

$$\frac{d\mathbf{P}}{dt} = \sum \frac{d\mathbf{p}}{dt}, \tag{9.27}$$

the first relation (9.22) can also be written as

$$\frac{d\mathbf{P}}{dt} = \mathbf{F}. \tag{9.28}$$

It is worth mentioning that only the external forces are included in the resultant force $\mathbf{F}$ (the internal forces cancel each other).

To write the second relation (9.22) in a more convenient form, let us first observe that variation of the potential energy with respect to an infinitesimal rotation $\delta\boldsymbol{\alpha}$ of the body writes

$$\delta_\alpha U = \sum \frac{\partial U}{\partial \mathbf{r}} \delta\mathbf{r} = \sum \frac{\partial U}{\partial \mathbf{r}} (\delta\boldsymbol{\alpha} \times \mathbf{r})$$

$$= \sum \left(\mathbf{r} \times \frac{\partial U}{\partial \mathbf{r}} \right) \delta\boldsymbol{\alpha}, \tag{9.29}$$

so that

$$\sum \frac{d\mathbf{L}}{dt} = -\sum \left(\mathbf{r} \times \frac{\partial U}{\partial \mathbf{r}} \right). \tag{9.30}$$

This result can still be simplified by choosing a fixed frame in which, at the considered moment of time, the centre of inertia of the rigid body is at rest ($\mathbf{V} = 0$). As a result, the infinitesimal translation $\delta\mathbf{r}$ is substituted by the infinitesimal rotation

$$\delta\mathbf{r} \rightarrow \delta\mathbf{r}', \tag{9.31}$$

and (9.30) becomes

$$\sum \frac{d\mathbf{L}}{dt} = -\sum \left(\mathbf{r}' \times \frac{\partial U}{\partial \mathbf{r}'} \right). \tag{9.32}$$

or, in view of (9.26),

$$\sum \frac{d\mathbf{L}}{dt} = \sum \mathbf{r}' \times \mathbf{F} = \sum \mathbf{M}, \tag{9.33}$$

where

$$\mathbf{M} = \mathbf{r}' \times \mathbf{F} \tag{9.34}$$

stands for the moment of the force $\mathbf{F}$ with respect to the centre of inertia of the rigid body. By means of notations

$$\mathbf{L} = \sum \mathbf{L}; \quad \mathbf{M} = \sum \mathbf{M}, \tag{9.35}$$

we finally obtain

$$\frac{d\mathbf{L}}{dt} = \mathbf{M}. \tag{9.36}$$

Using (3.24), let us now determine the kinetic energy of the rigid body. Since

$$\boldsymbol{\Omega} \sum \delta \mathbf{L} = \boldsymbol{\Omega}\delta \sum \mathbf{L} = \Omega_i \delta(I_{ik}\Omega_k)$$

$$= \delta\left(\frac{1}{2}I_{ik}\Omega_i\Omega_k\right),$$

and

$$\mathbf{V} \sum \delta\mathbf{p} = \mathbf{V}\delta\left(\sum \mathbf{p}\right) = \mathbf{V}\delta\left(\sum m\mathbf{V}\right)$$

$$= \delta\left(\sum \frac{m}{2}\mathbf{V}^2\right) = \delta\left(\frac{\mu}{2}\mathbf{V}^2\right), \tag{9.37}$$

where

$$\mu = \sum m, \tag{9.38}$$

the kinetic energy of the rigid body finally writes

$$T = \frac{1}{2}\mu V^2 + \frac{1}{2}I_{ik}\Omega_i\Omega_k. \tag{9.39}$$

The first term represents the translational kinetic energy, as if the whole mass of the body were concentrated in its centre of inertia, while the second term defines the rotational energy, due to rotation of the body with angular velocity $\mathbf{\Omega}$ about an axis passing through its centre of inertia.

9.5. Homogeneous and isotropic rigid body

Let us characterize the internal structure of the homogeneous and isotropic rigid body by the functional dependence

$$H = H(\beta_1, \beta_2) \tag{9.40}$$

where β_1 and β_2 are two Euclidean invariants defined as

$$\beta_1 = \frac{1}{2}\mathbf{p}^2; \quad \beta_2 = \frac{1}{2}\mathbf{L}^2, \tag{9.41}$$

while its motion is given by the 1-form

$$\omega'_s = \mathbf{p}d\mathbf{r} + \mathbf{L}d\boldsymbol{\alpha} - Hdt. \tag{9.42}$$

Postulate 9.2. *The inertial motion of the homogeneous and isotropic rigid body is given by the closeness of 1-form (9.42). It then follows*

$$D\omega'_s = d\mathbf{p} \wedge d\mathbf{r} + d\mathbf{L} \wedge d\boldsymbol{\alpha} - dH \wedge dt$$
$$= d\mathbf{p}\delta\mathbf{r} - \delta\mathbf{p}d\mathbf{r} + d\mathbf{L}\delta\boldsymbol{\alpha} - \delta\mathbf{L}d\boldsymbol{\alpha} - dH\delta t + \delta Hdt = 0. \tag{9.43}$$

Observing that

$$\delta H = \frac{\partial H}{\partial \beta_1}\mathbf{p}\delta\mathbf{p} + \frac{\partial H}{\partial \beta_2}\mathbf{L}\delta\mathbf{L} \tag{9.44}$$

then multiplying (9.43) by $(dt)^{-1}$ and canceling the coefficients of the variations $\delta\mathbf{r}, \delta\boldsymbol{\alpha}, \delta t$, we are left with:

a) The momentum conservation law

$$\frac{d\mathbf{p}}{dt} = 0; \quad \mathbf{p} = const. \tag{9.45}$$

b) The angular momentum conservation law

$$\frac{d\mathbf{L}}{dt} = 0; \quad \mathbf{L} = const. \tag{9.46}$$

c) The energy conservation law

$$\frac{dH}{dt} = 0; \quad H = const. \tag{9.47}$$

d) Relation of definition of linear momentum

$$\mathbf{p} = m\mathbf{v}, \tag{9.48}$$

where the mass m is given by

$$m = \left(\frac{\partial H}{\partial \beta_1}\right)^{-1}. \tag{9.49}$$

e) Relation expressing connection between angular momentum $\mathbf{L}$ and angular velocity $\mathbf{\Omega}$

$$\mathbf{L} = I\mathbf{\Omega}, \tag{9.50}$$

where the moment of inertia I is given by

$$\left(\frac{\partial H}{\partial \beta_2}\right)^{-1} = I. \tag{9.51}$$

Formula (9.50) follows from (9.15) for

$$I_{il} = I\delta_{il}, \tag{9.52}$$

which corresponds to a spherical rigid body. Such a body behaves like a material point.

f) Relation connecting the variations $\delta\mathbf{p}$, $\delta\mathbf{L}$, and δH

$$-\mathbf{v}\delta\mathbf{p} - \mathbf{\Omega}\delta\mathbf{L} + \delta H = 0. \tag{9.53}$$

This relation can be conveniently transformed by admitting the following simplifications:

i) Any shape modification of the rigid body, due to its rotation, is neglected. In other words, regardless of the state of motion of the rigid body, the ratio between its moment of inertia and its mass is a constant (**Postulate 9.3**), that is

$$\frac{I}{m} = \frac{I_0}{m_9} = const.$$

ii) Since the rigid body (a rigid sphere) is assimilable with a material point, the ratio $\frac{H}{m}$ has to be a constant (**Postulate 9.4**):

$$\frac{H}{m} = \varepsilon\omega^2; \quad \varepsilon = \pm 1.$$

Under these assumptions, relation (9.53) can be written as

$$\delta(\mathbf{p}^2 + m_0 I_0^{-1}\mathbf{L}^2 - \varepsilon\omega^2 m^2) = 0,$$

and, by integrating,

$$\mathbf{p}^2 + m_0 I_0^{-1}\mathbf{L}^2 - \varepsilon\omega^2 m^2 = -\varepsilon\omega^2 m_0^2 + + m_0 I_0^{-1}\mathbf{L}_0$$

Multiplying this relation by $(-\varepsilon\omega^2)$, we are left with

$$H^2 - \varepsilon\omega^2\mathbf{p}^2 - \mu^2\mathbf{L}^2 = m_0^2\omega^4 - \mu^2 L_0^2, \tag{9.54}$$

where

$$\mu = \left(\varepsilon\omega^2\frac{m_0}{I_0}\right)^{\frac{1}{2}}. \tag{9.55}$$

Here m_0 is the "rest mass", i.e. mass corresponding to $\mathbf{V} = 0$ (no translation of the centre of inertia) and $\mathbf{\Omega} = 0$ (no rotation).

Equation (9.54) involves some interesting peculiar cases. For example, the choices $\mathbf{L} = 0$, $\mathbf{L}_0 = 0$, $\varepsilon = +1$, and $\omega = c$ produce the well-known relation

$$H^2 = \mathbf{p}^2 c^2 + m_0^2 c^4.$$

Taking $\mathbf{p} = 0$, $\mathbf{L}_0 = 0$, $\varepsilon = +1$, and $\omega = c$, one obtains

$$H^2 = \mu^2\mathbf{L}^2 + \mu^4 I_0^2.$$

9.6. Rigid body with continuous mass distribution

Let us go back to relation (9.15). Its inverse writes

$$\Omega_l = \sigma_{lm}L_m, \tag{9.56}$$

with

$$\sigma_{lm} = I_{lm}^{-1}. \tag{9.57}$$

If

$$\Delta = \det(I_{lm}) = I_{11}I_{22}I_{33} - I_{11}I_{23}^2$$
$$-I_{22}I_{13}^2 - I_{33}I_{12}^2 + 2I_{12}I_{23}I_{13} \neq 0, \tag{9.58}$$

equations (9.56) can be written as

$$\begin{aligned}
\Omega_1 &= \chi_1 L_1 - \lambda_3 L_2 + \lambda_2 L_3; \\
\Omega_2 &= -\lambda_3 L_1 + \chi_2 L_2 - \lambda_1 L_3; \\
\Omega_3 &= -\lambda_2 L_1 + \lambda_1 L_2 + \chi_3 L_3,
\end{aligned} \tag{9.59}$$

where

$$\begin{aligned}
\chi_1 &= (I_{22}I_{33} - I_{23}^2)\Delta^{-1}; & \lambda_1 &= (I_{11}I_{23} - I_{12}I_{13}\Delta^{-1}; \\
\chi_2 &= (I_{11}I_{33} - I_{13}^2)\Delta^{-1}; & \lambda_2 &= (-I_{22}I_{13} + I_{12}I_{23})\Delta^{-1}; \\
\chi_3 &= (I_{11}I_{22} - I_{12}^2)\Delta^{-1}; & \lambda_3 &= (I_{33}I_{12} - I_{13}I_{23})\Delta^{-1}.
\end{aligned} \tag{9.60}$$

As one can see, the differential

$$\Omega_l \delta\Omega_l = \delta\left[\frac{1}{2}(\chi_l L_l^2)\right] + \varepsilon_{lkm}\lambda_l L_k \delta L_m, \tag{9.61}$$

where ε_{lkm} is the Levi-Civita symbol, puts into evidence both holonomic

$$\omega_i = \frac{1}{2}L_i^2 \tag{9.62}$$

and non-holonomic

$$\delta\varphi_l = \varepsilon_{lkm}L_k \delta L_m \tag{9.63}$$

invariants. Consequently, the Hamiltonian H is a functional of invariants

$$H = H(\beta_1, \omega_l, \varphi_l), \tag{9.64}$$

and we can write

$$\delta H = \frac{\partial H}{\partial \beta_1}\mathbf{p}\delta\mathbf{p} + \frac{\partial H}{\partial \omega_l}\delta\omega_l + \frac{\partial H}{\partial \varphi_l}\delta\varphi_l. \tag{9.65}$$

The principle of motion then takes the form

$$D\omega = d\mathbf{p}\delta\mathbf{r} + \delta\mathbf{p}d\mathbf{r} + d\mathbf{L}\delta\boldsymbol{\alpha} - \delta\mathbf{L}d\boldsymbol{\alpha} - dH\delta t$$

$$+ \left(\frac{\partial H}{\partial \beta_1}\mathbf{p}\delta\mathbf{p} + \frac{\partial H}{\partial \omega_l}\delta\omega_l + \frac{\partial H}{\partial \varphi_l}\delta\varphi_l \right) dt = 0. \qquad (9.61)$$

Multiplying (9.66) by $(dt)^{-1}$, then canceling the coefficients of variations $\delta\mathbf{r}, \delta\boldsymbol{\alpha}, \delta t, \delta\mathbf{p}$, equations (9.45), (9.46), (9.47), and (9.48) are retrieved. The only difference from the case of isotropic and homogeneous rigid body appears in canceling of the coefficient of variation $\delta\mathbf{L}$, due to the presence of non-holonomic invariants (9.63). Therefore, appears as necessary the identification

$$\Omega_l\delta L_l = \chi_l\delta\Omega_l + \lambda_l\delta\varphi_l = \left(\frac{\partial H}{\partial \omega_l}\delta\omega_l + \frac{\partial H}{\partial \varphi_l}\delta\varphi_l \right), \qquad (9.67)$$

which yields

$$\chi_l = \frac{\partial H}{\partial \omega_l}; \qquad \lambda_l = \frac{\partial H}{\partial \varphi_l}, \qquad (9.68)$$

showing that the structure coefficients of a rigid body are dependent on the mass distribution.

Next, let us suppose that the tensor

$$\varepsilon_{lm} = \frac{I_{lm}}{m} \qquad (9.69)$$

with dimensions of a squared length, if the case of the classic homogeneous rigid body is considered, does not depend on the mass distribution. Then, by inverting relations

$$L_l = m\varepsilon_{lm}\Omega_m, \qquad (9.70)$$

one obtains:

$$\begin{cases} m\Omega_1 = u_1 L_1 - v_3 L_2 + v_2 L_3, \\ m\Omega_2 = v_3 L_1 + u_2 L_2 - v_1 L_3, \\ m\Omega_3 = -v_2 L_1 + v_1 L_2 + u_3 L_3, \end{cases} \qquad (9.71)$$

where

$$\begin{cases} u_1 = (\varepsilon_{22}\varepsilon_{33} - \varepsilon_{23}^2)\Delta'^{-1}; \\ u_2 = (\varepsilon_{11}\varepsilon_{33} - \varepsilon_{13}^2)\Delta'^{-1}; \\ u_3 = (\varepsilon_{11}\varepsilon_{22} - \varepsilon_{12}^2)\Delta'^{-1}; \end{cases} \quad \begin{cases} v_1 = (\varepsilon_{11}\varepsilon_{23} - \varepsilon_{12}\varepsilon_{13})\Delta'^{-1}; \\ v_2 = (-\varepsilon_{22}\varepsilon_{13} + \varepsilon_{12}\varepsilon_{23})\Delta'^{-1}; \\ v_3 = (\varepsilon_{33}\varepsilon_{12} - \varepsilon_{13}\varepsilon_{23})\Delta'^{-1}, \end{cases}$$

$$(9.72)$$

and

$$\Delta' = \varepsilon_{11}\varepsilon_{22}\varepsilon_{33} - \varepsilon_{11}\varepsilon_{23}^2$$

$$-\varepsilon_{22}\varepsilon_{13}^2 - \varepsilon_{33}\varepsilon_{12}^2 + 2\varepsilon_{12}\varepsilon_{12}\varepsilon_{23} \neq 0. \tag{9.73}$$

The quantities ε_{ik} depend only on spatial structure of the rigid body.

Postulate 9.3. *The ratio between energy and inertial mass of a the rigid body is a constant, that is*

$$\frac{H}{m} = \omega^2. \tag{9.74}$$

Multiplying (9.66) by m and using (9.74), we obtain

$$-\mathbf{p}\delta\mathbf{p} - m\,\mathbf{\Omega}\,\delta\mathbf{L} + \omega^2 m\delta m = 0. \tag{9.75}$$

Since

$$m\,\mathbf{\Omega}\,\delta\mathbf{L} = \delta\left(\frac{1}{2}u_l L_l^2 + v_l\varphi_l\right), \tag{9.76}$$

we still have

$$\delta\left(\frac{1}{2}\mathbf{p}^2 + \frac{1}{2}u_l L_l^2 + v_l\varphi_l\right) = \delta\left(\frac{1}{2}\omega^2 m^2\right), \tag{9.77}$$

and, by integration

$$\mathbf{p}^2 + u_l L_l^2 + 2v_l\varphi_l = \omega^2(m^2 - m_0^2), \tag{9.78}$$

where m_0 is the rest mass of the rigid body for

$$\mathbf{v} \to 0,\,; \qquad \mathbf{\Omega} \to 0. \tag{9.79}$$

The final results then write:

$$m = [m_0^2 + \omega^{-2}(\mathbf{p}^2 + u_l L_l^2 + 2v_l\varphi_l)]^{\frac{1}{2}}, \qquad (9.80)$$

and

$$H = \omega^2[m_0^2 + \omega^{-2}(\mathbf{p}^2 + u_l L_l^2 + 2v_l\varphi_l)]^{\frac{1}{2}}. \qquad (9.81)$$

9.7. Eulerian rigid body

The motion of a rigid body can be described by three coordinates of its centre of inertia, together with Euler's angles θ, φ, ψ.

The components of the angular velocity $\mathbf{\Omega}$ are given by

$$\begin{aligned}
\Omega_1 &= \dot{\varphi}\sin\theta\sin\psi + \dot{\theta}\cos\psi, \\
\Omega_2 &= \dot{\varphi}\sin\theta\cos\psi - \dot{\theta}\sin\psi, \\
\Omega_3 &= \dot{\varphi}\cos\theta + \dot{\psi}.
\end{aligned} \qquad (9.82)$$

Relations (9.82) can be written in terms of a single parameter τ signifying the time. In this respect, one integrate Euler's equations

$$\begin{aligned}
I_1\frac{d\Omega_1}{dt} + (I_3 - I_2)\Omega_2\Omega_3 &= 0, \\
I_2\frac{d\Omega_2}{dt} + (I_1 - I_3)\Omega_3\Omega_1 &= 0, \\
I_3\frac{d\Omega_3}{dt} + (I_2 - I_1)\Omega_1\Omega_2 &= 0,
\end{aligned} \qquad (9.83)$$

using the laws of conservation of energy

$$I_1\Omega_1^2 + I_2\Omega_2^2 + I_3\Omega_3^2 = 2E = const. \qquad (9.84)$$

and angular momentum

$$I_1^2\Omega_1^2 + I_2^2\Omega_2^2 + I_3^2\Omega_3^2 = L^2 = const., \qquad (9.85)$$

where

$$I_1 = I_{11}; \quad I_2 = I_{22}; \quad I_3 = I_{33}. \qquad (9.86)$$

The result is

$$\Omega_1 = \left[\frac{2EI_3 - L^2}{I_1(I_3 - I_1)}\right]^{\frac{1}{2}} cn\tau,$$

$$\Omega_2 = \left[\frac{2EI_3 - L^2}{I_2(I_3 - I_2)}\right]^{\frac{1}{2}} sn\tau,$$

$$\Omega_3 = \left[\frac{L^2 - 2EI_1}{I_3(I_3 - I_1)}\right]^{\frac{1}{2}} dn\tau. \tag{9.87}$$

Here

$$\tau = t \left[\frac{(I_3 - I_2)(L^2 - 2EI_1)}{I_1 I_2 I_3}\right]^{\frac{1}{2}}, \tag{9.88}$$

while cn, sn, dn are the Jacobi elliptic functions [35], with period T

$$T = 4k \left[\frac{I_1 I_2 I_3}{(I_3 - I_2)(L^2 - 2EI_1)}\right]^{\frac{1}{2}} F\left(\frac{\pi}{2}, k\right) \tag{9.89}$$

where

$$F\left(\frac{\pi}{2}, k\right) = \int_0^{\pi/2} \frac{d\varphi}{(1 - k^2 \sin^2 \varphi)^{1/2}} \tag{9.90}$$

and

$$k^2 = \frac{(I_2 - I_1)(2EI_3 - L^2)}{(I_3 - I_2)(L^2 - 2EI_1)}. \tag{9.91}$$

Using (9.80), the mass of rigid body finally writes

$$m = [m_0^2 + \omega^{-2}(I_1\Omega_1^2 + I_2\Omega_2^2 + I_3\Omega_3^2)]^{1/2}$$

$$= \left[m_0^2 + \omega^{-2}\left(\frac{2EI_3 - L^2}{I_3 - I_1}cn^2\tau + \frac{2EI_3 - L^2}{I_3 - I_2}sn^2\tau\right.\right.$$

$$\left.\left. + \frac{L^2 - 2EI_1}{I_3 - I_1}dn^2\tau\right)\right]^{1/2}. \tag{9.92}$$

As one can see, the mass is an elliptic function of time.

9.8. Motion in a field of the rigid body

Let us substitute roto-translation (9,3) in the field 1-form $\omega = \mathbf{A}d\mathbf{r} - A_0 dt$. The result is

$$\omega = \mathbf{A}d\mathbf{R} + \mathbf{B}d\boldsymbol{\alpha} - A_0 dt, \qquad (9.93)$$

where

$$\mathbf{B} = \mathbf{r} \times \mathbf{A}. \qquad (9.94)$$

Consequently, in case of rigid displacements, the field is defined by means of two vector potentials $\mathbf{A}, \mathbf{B}$, and one scalar potential A_0. Quantities $\mathbf{A}, \mathbf{B}, A_0$ must be apriorically considered as functions of $(\mathbf{r}, \boldsymbol{\alpha}, t)$.

The exterior differential of the 1-form (9.93) is

$$D\omega = d\mathbf{A} \wedge d\mathbf{R} + d\mathbf{B} \wedge d\boldsymbol{\alpha} - dA_0 \wedge dt$$

$$= d\mathbf{A}\delta\mathbf{R} - \delta\mathbf{A}d\mathbf{R} + d\mathbf{B}\delta\boldsymbol{\alpha} - \delta\mathbf{B}d\boldsymbol{\alpha} - dA_0\delta t + \delta A_0 dt$$

$$= \left(dA_i - \frac{\partial A_l}{\partial x_i}dx_l - \frac{\partial B_l}{\partial x_i}d\alpha_l + \frac{\partial A_0}{\partial x_i}dt \right) \delta x_i \qquad (9.95)$$

$$+ \left(dB_i - \frac{\partial A_l}{\partial x_i}dx_l - \frac{\partial B_l}{\partial x_i}d\alpha_l + \frac{\partial A_0}{\partial \alpha_i}dt \right) \delta \alpha_i$$

$$- \left(dA_0 + \frac{\partial A_l}{\partial t}dx_l + \frac{\partial B_l}{\partial t}d\alpha_l + \frac{\partial A_0}{\partial t}dt \right) \delta t.$$

Postulate 9.4. *The motion of the rigid body in the set of fields* $(\mathbf{A}, \mathbf{B}, A_0)$ *is given by equality of exterior differentials of 1-forms (9.42) and (9.93).*

Since

$$D\omega = dp_i\delta x_i + dL_i\delta\alpha_i - dH\delta t$$

$$- \left(dx_i - \frac{\partial H}{\partial p_i}dt \right) \delta p_i - \left(d\alpha_i - \frac{\partial H}{\partial L_i}dt \right) \delta L_i, \qquad (9.96)$$

equating relations (9.95) and (9.96), then multiplying the result by $(dt)^{-1}$, we obtain:

a) Relations of definition for velocities

$$\frac{dx_i}{dt} = \frac{\partial H}{\partial p_i}; \qquad \frac{d\alpha_i}{dt} = \frac{\partial H}{\partial L_i}; \tag{9.97}$$

b) Equations of motion

$$\frac{dp_i}{dt} = \frac{dA_i}{dt} - \frac{\partial A_l}{\partial x_i}\dot{x}_l - \frac{\partial B_l}{\partial x_i}\dot{\alpha}_l + \frac{\partial A_0}{\partial x_i}; \tag{9.98}$$

$$\frac{dL_i}{dt} = \frac{dB_i}{dt} - \frac{\partial A_l}{\partial \alpha_i}\dot{x}_l - \frac{\partial B_l}{\partial \alpha_i}\dot{\alpha}_l + \frac{\partial A_0}{\partial \alpha_i}. \tag{9.99}$$

c) Balance equation of energy exchange between the rigid body and the field

$$\frac{dH}{dt} = \frac{dA_0}{dt} - \frac{\partial A_0}{\partial t} + \frac{\partial A_l}{\partial t}\dot{x}_l + \frac{\partial B_l}{\partial t}\dot{\alpha}_l. \tag{9.100}$$

Equation (9.98) can also be written as

$$\frac{dp_i}{dt} = \frac{\partial A_i}{\partial t} - \left(\frac{\partial A_i}{\partial x_l} - \frac{\partial A_l}{\partial x_i}\right)\dot{x}_l$$
$$- \left(\frac{\partial A_i}{\partial \alpha_l} - \frac{\partial B_l}{\partial x_i}\right)\dot{\alpha}_l + \frac{\partial A_0}{\partial x_i}, \tag{9.101}$$

where we used

$$\frac{dA_i}{dt} = \frac{\partial A_i}{\partial t} + \frac{\partial A_i}{\partial x_l}\dot{x}_l + \frac{\partial A_i}{\partial \alpha_l}\dot{\alpha}_l. \tag{9.102}$$

If the fields set $(\mathbf{A}, \mathbf{B}, A_0)$ are functions of (x_l, t) only, then (9.98), (9.99), and (9.100) become:

$$\frac{d\mathbf{p}}{dt} = -(\mathbf{E} + \mathbf{v} \times \mathbf{B}) - \nabla(\dot{\boldsymbol{\alpha}} \cdot \mathbf{B}); \tag{9.103}$$

$$\frac{d\mathbf{L}}{dt} = \frac{d\mathbf{B}}{dt}; \tag{9.104}$$

$$\frac{dH}{dt} = -\mathbf{E} \cdot \mathbf{v} + \frac{\partial}{\partial t}(\mathbf{B} \cdot (\dot{\boldsymbol{\alpha}}), \tag{9.105}$$

where

$$\mathbf{E} = \frac{\partial \mathbf{A}}{\partial t} + \nabla A_0; \tag{9.106}$$

$$\mathbf{B} = \nabla \times \mathbf{A}. \tag{9.107}$$

9.9. Calculation of invariantive potentials

In case of a continuous mass (electric charge, etc.) distribution, to determine the field quantities one usually consider elementary 1-forms of the type

$$\omega = \mathbf{a}(\boldsymbol{\xi}, t)d\boldsymbol{\xi} - a_0(\boldsymbol{\xi}, t)\, dt, \tag{9.108}$$

corresponding to the infinitesimal components of mass $d\mu$ of the body. By definition, one takes

$$\omega = \int_{\text{body}} [\mathbf{a}(\boldsymbol{\xi}, t)d\boldsymbol{\xi} - a_0(\boldsymbol{\xi}, t)dt]\, d\mu, \tag{9.109}$$

with implicit hypothesis that the mass

$$\mu = \int d\mu \tag{9.110}$$

is finite. The integral (9.110) is only considered over the body space.

If the rigid displacement (9.3) is considered, relation (9.109) becomes

$$\omega = \int \{\mathbf{a}(\boldsymbol{\xi}, t)d\mathbf{R} + [\mathbf{r} \times \mathbf{a}(\boldsymbol{\xi}, t)]d\boldsymbol{\alpha} - \alpha_0(\boldsymbol{\xi}, t)dt\}\, d\mu, \tag{9.111}$$

or

$$\omega = \mathbf{A}(\mathbf{R}, t)d\mathbf{R} + \mathbf{B}(\mathbf{R}, t)d\alpha - A_0(\mathbf{R}, t)dt, \tag{9.112}$$

where the following notations have been introduced:

$$\mathbf{A}(\mathbf{R}, t) = \int \mathbf{a}(\boldsymbol{\xi}, t)d\mu; \tag{9.113}$$

$$\mathbf{B}(\mathbf{R}, t) = \int \mathbf{b}(\boldsymbol{\xi}, t)d\mu = \int [\mathbf{r} \times \mathbf{a}(\boldsymbol{\xi}, t)]d\mu; \tag{9.114}$$

$$A_0(\mathbf{R}, t) = \int a_9(\boldsymbol{\xi}, t) \, d\mu. \tag{9.115}$$

Here we used the fact that the integrals (9.113), (9.114), and (9.115) do not depend on $\mathbf{R}$ and t; they are calculated with respect to the centre of inertia frame.

9.10. Space of corpuscles

Presuming that corpuscle dimensions allow only simple approximations, let us suppose that functions $\mathbf{a}(\boldsymbol{\xi}, t)$ admit continuous partial derivatives up to the second order. To represent the radius vector $\mathbf{r}$, we choose a frame connected to the rigid body. Denoting by γ_i $(i = \overline{1,3})$ the components of $\mathbf{r}$, we have:

$$\mathbf{a}(\boldsymbol{\xi}, t) = \mathbf{a}(\mathbf{R}, t) + \sum_l \gamma_l \frac{\partial \mathbf{a}}{\partial R_l}(\mathbf{R}, t) + \frac{1}{2} \sum_{l,n} \gamma_l \gamma_n \frac{\partial^2 \mathbf{a}}{\partial R_l \partial R_n}$$

$$+ \frac{1}{2} \sum_{l,n} \gamma_l \gamma_n \left[\frac{\partial^2 \mathbf{a}}{\partial R_l \partial R_n}(R_1 + u_1 \gamma_1, R_2 + u_2 \gamma_2, R_3 + u_3 \gamma_3, t) \right.$$

$$\left. - \frac{\partial^2 \mathbf{a}}{\partial R_l \partial R_n}(\mathbf{R}, t) \right]. \tag{9.116}$$

with

$$|u_1|, \ |u_2|, \ |u_3| < 1. \tag{9.117}$$

We then have:

$$\mathbf{A}(\mathbf{R}, t) = \mu \mathbf{a}(\boldsymbol{\xi}, t) + \sum_i l_i \frac{\partial \mathbf{a}}{\partial R_i}(\mathbf{R}, t) + \frac{1}{2} \sum_{i,j} l_{ij} \frac{\partial^2 \mathbf{a}}{\partial R_i \partial R_j}(\mathbf{R}, t) + \varepsilon, \tag{9.118}$$

with

$$l_i = \int \gamma_i \, d\mu; \qquad l_{ij} = \int \gamma_i \gamma_j \, d\mu, \tag{9.119}$$

and

$$\varepsilon = \frac{1}{2} \sum_{l,n} \gamma_l \gamma_n \left[\frac{\partial^2 \mathbf{a}}{\partial R_l \partial R_n} (R_1 + u_1 \gamma_1, R_2 + u_2 \gamma_2, R_3 + u_3 \gamma_3, t) \right.$$

$$\left. - \frac{\partial^2 \mathbf{a}}{\partial R_l \partial R_n} (\mathbf{R}, t) \right]. \tag{9.120}$$

Using the same conventions and (9.112) as well, we can write

$$\mathbf{B}(\mathbf{R}, t) = \int [\mathbf{r} \times \mathbf{a}(\boldsymbol{\xi}, t)] \, d\mu$$

$$= \boldsymbol{\ell} \times \mathbf{a}(\mathbf{R}, t) + \sum_i \left(\boldsymbol{\ell}_i \times \frac{\partial \mathbf{a}}{\partial R_i} \right) + \boldsymbol{\chi}, \tag{9.121}$$

where $\boldsymbol{\ell}$ is a vector of components $\int \gamma_i d\mu$, $\boldsymbol{\ell}_i$ is a set of vectors of components $\int \gamma_i \gamma_l \, d\mu$, while $\boldsymbol{\chi}$ means

$$\chi = \int \gamma_i \gamma_l \gamma_n b(\gamma, t) \, d\mu \tag{9.122}$$

According to our initial hypothesis, χ is supposed to be negligible. Therefore, we have:

$$\omega = \left(\mu \mathbf{a} + \sum_i \ell_i \frac{\partial \mathbf{a}}{\partial R_i} + \frac{1}{2} \sum_{i,l} \ell_{il} \frac{\partial^2 \mathbf{a}}{\partial R_i \partial R_l} + \varepsilon \right) d\mathbf{R}$$

$$+ \left[\boldsymbol{\ell} \times \mathbf{a} + \sum_i \left(\boldsymbol{\ell}_i \times \frac{\partial \mathbf{a}}{\partial R_i} \right) + \boldsymbol{\chi} \right] d\boldsymbol{\alpha}$$

$$- \left(\mu \, a_0 + \sum_i \ell_i \frac{\partial a_0}{\partial R_i} + \frac{1}{2} \sum_{i,l} \ell_{il} \frac{\partial^2 a_0}{\partial R_i \partial R_l} + \varepsilon_0 \right) dt, \tag{9.123}$$

with

$$\varepsilon_o = \frac{1}{2} \sum_{l,n} \gamma_l \gamma_n \left[\frac{\partial^2 a_0}{\partial R_l \partial R_n} (R_1 + u_1 \gamma_1, R_2 + u_2 \gamma_2, R_3 + u_3 \gamma_3, t) \right.$$

$$\left. - \frac{\partial^2 a_0}{\partial R_l \partial R_n} (\mathbf{R}, t) \right]. \tag{9.124}$$

The coefficients ℓ_i and ℓ_{ik} are invariant characteristics of the rigid body.

9.11. Classification of corpuscles

A particle is called *particle of the first species* if the terms containing ℓ_{ik} are negligible. In this case, the 1-form (9.123) writes

$$\omega = \left(\mu\mathbf{a} + \sum_i \ell_i \frac{\partial \mathbf{a}}{\partial R_i} \right) d\mathbf{R}$$

$$+ (\ell \times \mathbf{a}) d\boldsymbol{\alpha} - \left(\mu a_0 + \sum_i \ell_i \frac{\partial \mathbf{a}}{\partial R_i} \right) dt. \qquad (9.125)$$

A particle is *of the second species* if the terms containing higher powers of ℓ_{ik} are neglected. The corresponding 1-form therefore is

$$\omega = \left(\mu\mathbf{a} + \sum_i \frac{\partial \mathbf{a}}{\partial R_i} + \frac{1}{2} \sum_{i,k} \ell_{ik} \frac{\partial^2 \mathbf{a}}{\partial R_i \partial R_k} \right) d\mathbf{R}$$

$$+ \left[\ell \times \mathbf{a} + \sum_i \left(\ell_i \times \frac{\partial \mathbf{a}}{\partial R_i} \right) \right] d\boldsymbol{\alpha} \qquad (9.126)$$

$$- \left(\mu a_0 + \sum_i \ell_i \frac{\partial a_0}{\partial R_i} + \frac{1}{2} \sum_{i,k} \ell_{ik} \frac{\partial^2 a_0}{\partial R_i \partial R_k} \right) dt$$

The equations of motion of a corpuscle of the first species is then obtained from (9.123) by choosing

$$H_1 = (\ell \times \mathbf{a})\boldsymbol{\alpha}, \qquad (9.127)$$

$$\mathbf{L}_1 = \mathbf{L}_1^0 + \ell \times [\mathbf{a}(\mathbf{R}, t) - \mathbf{a}(\mathbf{R}, 0)]. \qquad (9.128)$$

For particles of the second species, a similar reasoning requires:

$$H_2 = \left[\ell \times \mathbf{a} + \sum_i \ell_i \times \frac{\partial \mathbf{a}}{\partial R_i} \right] \boldsymbol{\alpha}; \qquad (9.129)$$

$$\mathbf{L}_2 = \mathbf{L}^0 + \boldsymbol{\ell} \times [\mathbf{a}(\mathbf{R}, t) - \mathbf{a}(\mathbf{R}, 0)]$$

$$+ \sum_i \left\{ \boldsymbol{\ell}_i \times \left[\frac{\partial \mathbf{a}}{\partial R_i}(\mathbf{R}, t) - \frac{\partial \mathbf{a}}{\partial R_i}(\mathbf{R}, 0) \right] \right\}. \qquad (9.130)$$

Equations (9.121) and (9.125) have to be completed with (9.97), where H_1 is given by (9.127).

Chapter 10

Covariant Formulation of Conservation Laws in Invariantive Mechanics

10.1. Conservation laws for one material point

Let us consider the 1-form

$$\omega = \mathbf{p}d\mathbf{r} - Hdt \tag{10.1}$$

and use the infinitesimal Lorentz transformation

$$\Delta \mathbf{r} = d\mathbf{r} + td\mathbf{V} + (\mathbf{r} \times d\mathbf{\Omega});$$

$$\Delta t = dt + \omega^{-2}\mathbf{r}d\mathbf{V}. \tag{10.2}$$

The result is

$$\omega^i = \mathbf{p}d\mathbf{r} - Hdt + \mathbf{L}d\mathbf{\Omega} - \omega^{-1}\mathbf{K}d\mathbf{V}, \tag{10.3}$$

where $\mathbf{L}$ is the angular momentum

$$\mathbf{L} = \mathbf{r} \times \mathbf{p} \tag{10.4}$$

and

$$\mathbf{K} = \omega^{-1}H\mathbf{r} - \omega\mathbf{p}t. \tag{10.5}$$

Postulate 10.1. *The motion is described by the closeness of the 1-form (10.3), that is*

$$D\omega^i = 0. \tag{10.6}$$

147

It follows that

$$D\omega^i = d\mathbf{p} \wedge d\mathbf{r} - dH \wedge dt + d\mathbf{L} \wedge d\mathbf{\Omega} - \omega^{-1}d\mathbf{K} \wedge d\mathbf{V}$$
$$= (d\mathbf{p}\delta\mathbf{r} - \delta\mathbf{p}d\mathbf{r}) - (dH\delta t - \delta H dt)$$
$$+ (d\mathbf{L}\delta\mathbf{\Omega} - \delta\mathbf{L}d\mathbf{\Omega}) - \omega^{-1}(d\mathbf{K}\delta\mathbf{V} - \delta\mathbf{K}d\mathbf{V}). \qquad (10.7)$$

Since variations $\delta\mathbf{r}, \delta t, \delta\mathbf{\Omega}, \omega^{-1}\delta\mathbf{V}$ are arbitrary, their coefficients are equal to zero. Multiplying by $(dt)^{-1}$, one then obtains:

a) The linear momentum conservation law

$$\frac{d\mathbf{p}}{dt} = 0; \quad \mathbf{p} = const. \qquad (10.8)$$

b) The energy conservation law

$$\frac{dH}{dt} = 0; \quad H = const. \qquad (10.9)$$

c) The angular momentum conservation law

$$\frac{d\mathbf{L}}{dt} = 0; \quad \mathbf{L} = const. \qquad (10.10)$$

d) The conservation law

$$\frac{d\mathbf{K}}{dt} = 0; \quad \mathbf{K} = const., \qquad (10.11)$$

as well as the relation

$$-\mathbf{V}\delta\mathbf{p} + \delta H - \mathbf{\Omega}\delta\mathbf{L} + \omega^{-1}\boldsymbol{\alpha}\,\delta\mathbf{K} = 0. \qquad (10.12)$$

10.2. Equations of conservation for a discrete system of material points

If the system is composed by n material points, the 1-form (19.3) writes

$$\omega^i = \sum_{\ell=1}^{n}(\mathbf{p}_\ell d\mathbf{r}_\ell - H_\ell dt + \mathbf{L}_\ell d\mathbf{\Omega}_\ell - \omega^{-1}\mathbf{K}_\ell d\mathbf{V}_\ell). \qquad (10.13)$$

The closeness of this 1-form yields the following laws of conservation:

$$\frac{d\mathbf{P}}{dt} = \frac{d}{dt}\left(\sum_{\ell}\mathbf{p}_\ell\right) = 0;$$

$$\frac{dH}{dt} = \frac{d}{dt}\left(\sum_{\ell}H_\ell\right) = 0; \qquad (10.14)$$

$$\frac{d\mathbf{L}}{dt} = \frac{d}{dt}\left(\sum_{\ell}\mathbf{L}_\ell\right) = 0,$$

together with

$$\frac{d\mathbf{K}}{dt} = \frac{d}{dt}\sum_{\ell}\mathbf{K}_\ell = \frac{d}{dt}\left[\sum_{\ell}(\omega^{-1}H_\ell\mathbf{r}_\ell - \omega\mathbf{p}_\ell t)\right] = 0, \qquad (10.15)$$

as well as

$$\sum_{\ell}\left(-\mathbf{V}_\ell\delta\mathbf{p}_\ell + \delta H_\ell - \mathbf{\Omega}_\ell\delta\mathbf{L}_\ell + \omega^{-1}\boldsymbol{\alpha}_\ell\,\delta\mathbf{K}_\ell\right) = 0. \qquad (10.16)$$

Since the total angular momentum and total kinetic energy are conserved, equation (10.15) leads to

$$\frac{\sum_{\ell}H_\ell\mathbf{r}_\ell}{\sum_{\ell}H_\ell} - \omega^2\frac{\sum_{\ell}\mathbf{p}_\ell}{\sum_{\ell}H_\ell} = const. \qquad (10.17)$$

It then follows that the fictive particle of radius-vector

$$\mathbf{R} = \frac{\sum_{\ell}H_\ell\mathbf{r}_\ell}{\sum_{\ell}H_\ell} \qquad (10.18)$$

moves uniformly with constant velocity

$$\mathbf{v} = \omega^2\frac{\sum_{\ell}\mathbf{p}_\ell}{\sum_{\ell}H_\ell} = const. \qquad (10.19)$$

and momentum

$$\mathbf{P} = \omega^2 H\mathbf{V}. \qquad (01.20)$$

Relation (10.18) stands for the invariantive definition of the centre-of-mass.

10.3. Four-dimensional formulation of the conservation laws

Let us introduce the infinitesimal transformation

$$dx^\alpha \to dx^\alpha + x^\alpha d\Omega^\beta_\alpha \tag{10.21}$$

into the 1-form

$$\omega = p_\beta dx^\beta,$$

where

$$\Omega_{\alpha\beta} = -\Omega_{\beta\alpha} \tag{10.22}$$

is the antisymmetric tensor of 4-dimensional rotations

$$\Omega_{\alpha\beta} = \begin{bmatrix} 0 & \omega_z & -\omega_y & -i\omega^{-1}V_x \\ -\omega_z & 0 & \omega_x & -i\omega^{-1}V_y \\ \omega_y & -\omega_x & 0 & -i\omega^{-1}V_z \\ i\omega^{-1}V_x & i\omega^{-1}V_y & i\omega^{-1}V_z & 0 \end{bmatrix}. \tag{10.23}$$

This investigation leads to the generalized 1-form

$$\omega^i = p_\alpha dx^\alpha + \frac{1}{2} M^\alpha_\beta d\Omega^\beta_\alpha, \tag{10.24}$$

where

$$M^\alpha_\beta = x^\alpha p_\beta - x^\beta p_\alpha \tag{10.25}$$

is the angular momentum 4-tensor. Its components are:

$$M_{\alpha\beta} = \begin{bmatrix} 0 & L_z & -L_y & iK_x \\ -L_z & 0 & L_x & iK_y \\ L_y & -L_x & 0 & iK_z \\ -iK_x & -iK_y & -iK_z & 0 \end{bmatrix}. \tag{10.26}$$

Postulate 19.2. *The motion is described by the closeness of the 1-form (10.24).* One obtains:

$$D\omega^i = dp_\alpha \wedge dx^\alpha + \frac{1}{2} dM^\alpha_\beta \wedge d\Omega^\beta_\alpha$$

$$= dp_\alpha \delta x^\alpha - \delta p_\alpha dx^\alpha + \frac{1}{2}\left(dM^\alpha_\beta \delta\Omega^\beta_\alpha - \delta M^\alpha_\beta d\Omega^\beta_\alpha \right) = 0. \tag{10.27}$$

Therefore, we are left with:

a) The 4-momentum conservation law

$$\frac{dp_\alpha}{d\tau} = 0; \tag{10.28}$$

b) The law of conservation of angular momentum 4-tensor

$$\frac{dM_\beta^\alpha}{d\tau} = 0; \tag{10.29}$$

c) The operational relation

$$\delta p_\alpha u^\alpha + \frac{1}{2}\delta M_\beta^\alpha \dot{\Omega}_\alpha^\beta = 0. \tag{10.30}$$

This covariant formulation was used in [12] and [36] in order to study a fluid, in the absence and presence of the field, respectively.

Chapter 11

Invariantive Mechanics and Informational Energy

11.1. Preliminaries

The purpose of this chapter is to establish connection between Onicescu informational energy and the interaction constants, in case of proportionality of two 1-forms. Such a desiderate is possible due to connection between the group $SL(2R)$ and the canonical formalism.

11.2. The group $SL(2R)$ and the canonical formalism

The group $SL(2R)$ is the group of transformations

$$\begin{aligned}
x' &= \alpha x + \beta y; \\
y' &= \gamma x + \delta y; \\
\alpha\delta - \beta\gamma &= 1,
\end{aligned} \qquad (11.1)$$

which preserves areas in (x, y) space. Choosing parametrization as

$$\begin{aligned}
\alpha &= 1 + \frac{1}{2}a^2; \\
\beta &= a^1; \\
\gamma &= -a^3 \\
\delta &= 1 - \frac{1}{2}a^2,
\end{aligned} \qquad (11.2)$$

the infinitesimal group transformations become

$$x' = x + ya^1 + \frac{x}{2}a^2;$$

$$y' = y - \frac{y}{2}a^2 + xa^3,$$

(11.3)

so that the associated Lie algebra takes the form:

$$[L_1, L_2] = L_1; \quad [L_2, L_3] = L_3; \quad [L_2, L_1] = -2L_2, \tag{11.4}$$

where

$$L_1 = y\frac{\partial}{\partial x}; \quad L_2 = \frac{1}{2}\left(x\frac{\partial}{\partial x} - y\frac{\partial}{\partial y}\right); \quad L_3 = -x\frac{\partial}{\partial y} \tag{11.5}$$

are the vectors of the Lie basis.

The fact that the group is measurable, having the elementary measure given by the 2-form

$$\omega = dy \wedge dx. \tag{11.6}$$

where "$\wedge$" is the exterior product, shows that the group $SL(2R)$ is also simplectic. Consequently, the corresponding Hamiltonian dynamics is generated in the tangent space by means of the vectors $L_i(i = 1, 2, 3)$ or, more general. by the linear combination

$$L = cL_1 + 2bL_2 + aL_3; \quad a, b, c = const., \tag{11.7}$$

while the Hamiltonian follows as an invariant function along the trajectories tangent to vector (11.7), that is as solution of the equation

$$LH = 0. \tag{11.8}$$

In view of (11.5), the last equation writes

$$(bx + cy)\frac{\partial H}{\partial x} - (ax + by)\frac{\partial H}{\partial y} = 0. \tag{11.9}$$

Therefore, the characteristic differential system

$$\frac{dx}{bx + cy} = \frac{dy}{ax + by} = dt \tag{11.10}$$

admits the first integral

$$H(x, y) = \frac{1}{2}(ax^2 + 2bxy + cy^2). \qquad (11.11)$$

As we can observe, the system of differential equations (11.10) is the Hamilton's system

$$\dot{x} = \frac{\partial H}{\partial y}; \qquad \dot{y} = -\frac{\partial H}{\partial x} \qquad (11.12)$$

associated with the Hamiltonian (11.11). Formally, this is equivalent to the equations of motion of two linear harmonic oscillators of coordinates x and y

$$\ddot{x} + \Omega x = 0;$$
$$\ddot{y} + \Omega y = 0, \qquad (11.13)$$

with

$$\Omega = ac - b^2 > 0. \qquad (11.14)$$

Another first integral of equation (11.8) is the unnormed Gaussian

$$\rho(x, y) = A \exp[-H(x, y)]; \quad A = const. \qquad (11.15)$$

If

$$a > 0, \quad c > 0, \quad \Omega = ac - b^2 > 0, \qquad (11.16)$$

the solutions of the system (11.10), which are the functions $x(t)$ and $y(t)$ satisfying relations

$$x' = \left(\cos \Omega t + \frac{b}{\Omega} \sin \Omega t \right) x + \frac{c}{\Omega}(\sin \Omega t)y;$$
$$y' = -\frac{a}{\Omega}(\sin \Omega t)x + \left(\cos \Omega t - \frac{b}{\Omega} \sin \Omega t \right) y \qquad (11.17)$$

give the trajectories in (x, y) space for $H = const$. Obviously, another choice of the parameters (a, b, c) leads to some other solutions.

Let us suppose the identifications

$$x \equiv q; \qquad y \equiv p, \tag{11.18}$$

where q is the generalized coordinate, and p the generalized momentum. Then the 2-form (12.6)

$$\omega = dp \wedge dq \tag{11, 19}$$

signifies the area element, transformations (11.1) become canonical due to the current conservation $\iint dp \wedge dq$ — Liouville's theorem, while (11.15) can be conceived as momentum-coordinate probability density. Since both the Hamiltonian (11.11) and the Gaussian (11.15) are invariant group functions, it is not necessary to use the ergodic theorem, and the probability stands for a function of energy

$$\rho = \rho(H), \tag{11.20}$$

on the condition that $H(x, y)$ is not only generator of the motion, but also signifies energy.

If $\rho(x, y) \equiv \rho(q, p)$, the set of parameters (a, b, c) receive a statistical significance.

11.3. Informational entropy. Transitivity varieties

Consider the probability density $\rho(p, q)$, together with

$$\bar{q} = \iint q\rho(p, q)dpdq;$$

$$\bar{p} = \iint p\rho(p, q)dpdq;$$

$$(\delta q)^2 = \iint (q - \bar{q})^2 \rho(p, q)dpdq; \tag{11.21}$$

$$(\delta p)^2 = \iint (p - \bar{p})^2 \rho(p, q)dpdq;$$

$$cov(p, q) = \iint (q - \bar{q})(p - \bar{p})\rho(p, q)dpdq,$$

where $\bar{q}, \bar{p}$ are the mean values of p, q, $(\delta q)^2, (\delta p)^2$ — the mean quadratic deviations and $cov(p, q)$ — the variables covariance. Then, by minimizing Shannon's information entropy

$$H = - \iint \rho(p,q) \ln[\rho(p,q)] dp dq, \qquad (11.22)$$

we obtain the normed Gaussian

$$\rho(p,q) = \frac{\sqrt{ac - b^2}}{2\pi} \exp[-H(p,q)] \qquad (11.23)$$

where $H(p,q)$ is the quadratic form

$$H(p,q) = \frac{1}{2}[a(q - \bar{q})^2 + 2b(q - \bar{q})(p - \bar{p}) + c(p - \bar{p})^2] \qquad (11.24)$$

with

$$a = \frac{(\delta q)^2}{D}; \qquad b = -\frac{cov(p,q)}{D}; \qquad c = \frac{(\delta p)^2}{D} \qquad (11.25)$$

and

$$D = (\delta p)^2 (\delta q)^2 - cov^2(p,q). \qquad (11.26)$$

Consequently, any statistical hypothesis is specified by a particular choice of coefficients (a, b, c) of the squared form (11.24), the class of all hypotheses being given by the condition

$$H(p', q') = H(p,q), \qquad (11.27)$$

where

$$H(p', q') = \frac{1}{2}[a'(q' - \bar{q})^2 + 2b'(q' - \bar{q})(p' - \bar{p}) + c'(p' - \bar{p})^2] \qquad (11.28)$$

But the correspondence

$$(p', q') \rightarrow (p, q) \qquad (11.29)$$

through the unimodular transformation (11.1), implies the three-parameter group

$$a' = \delta^2 a - 2\gamma\delta b + \gamma^2 c;$$
$$b' = -\beta\delta a - (\beta\gamma + \alpha\delta)b - \alpha\gamma c; \qquad (11.30)$$
$$c' = \beta^2 a - 2\alpha\beta b + \alpha^2 c.$$

This group, with the same parametrization (11.2), admits infinitesimal transformations such as

$$a' = a - aa^2 + 2ba^3;$$
$$b' = b - aa^1 + ca^3; \qquad (11.31)$$
$$c' = c - 2ba^1 + ca^2.$$

Considered as a linear system in unknowns (a^1, a^2, a^3), the system (11.31) is incompatible. Therefore, a correlation between triplets

$$(a, b.c) \to (a', b', c') \qquad (11.32)$$

does not exist, which means that the action of the group in the space of these parameters is non-transitive. It then follows the functional dependence

$$F = F(a, b, c) \qquad (11.33)$$

whose invariance defines the transition varieties.

Explicitly, (11.33) is of the form

$$F = ac - b^2 \qquad (11.34)$$

being obtained as solution of the system

$$A_1\phi = -a\frac{\partial\phi}{\partial a} - 2b\frac{\partial\phi}{\partial c} = 0;$$
$$A_2\phi = -a\frac{\partial\phi}{\partial a} + c\frac{\partial\phi}{\partial c} = 0; \qquad (11.35)$$
$$A_3\phi = 2b\frac{\partial\phi}{\partial b} + c\frac{\partial\phi}{\partial b} = 0,$$

where A_1, A_2, A_3 are the infinitesimal generators of the group (11.30) and satisfy the Lie algebra

$$[A_1, A_2] = A_1; \qquad [A_2, A_3] = A_3; \qquad [A_3, A_1] = -2A_2. \qquad (11.36)$$

But the transitivity varieties are given by

$$F = ac - b^2 = const. \qquad (11.37)$$

Assuming that $H(p,q)$ signifies energy, condition (11.37) shows that a representative point of the space (p,q) displacing on a surface of constant energy, is also located on a surface of constant probability density (ergodic condition)

$$\frac{\sqrt{a'c' - b'^2}}{2\pi} e^{-H(p',q')} = \frac{\sqrt{ac - b^2}}{2\pi} e^{-H(p,q)}. \qquad (11.38)$$

To conclude, the class of statistical hypotheses specifically to Gaussians of the same mean value is given by the ergodic condition.

11.4. Onicescu informational energy and uncertainty relations

We define the Onicescu informational energy, corresponding to distribution $\rho(p,q)$, by

$$E = \iint_{-\infty}^{+\infty} \rho^2(p,q) dp dq. \qquad (11.39)$$

If $\rho(p,q)$ is the normed Gaussian (11.23) and $H(p,q) > 0$, definition (11.39) becomes

$$E(a,b.c) = \frac{\sqrt{ac - b^2}}{4\pi}. \qquad (11.40)$$

We observe that:

i) The quantity

$$A = \frac{1}{2E(a,b,c)} \qquad (11.41)$$

is a measure of areas of ellipses with $H(p,q) = const.$ and becomes an indicator of dispersal of distribution $\rho(p,q)$, so that the bigger the informational energy of Gaussian (11.23) is, the more grouped it behaves.

ii) If we define the principle of stationary informational energy by

$$E(a, b, c) = const. \qquad (11.42)$$

then

ii-1) The class of statistical hypotheses, specific to Gaussians of the same mean value, is given by the constant informational energy;

ii-2) A system is quasi-ergodic if its informational energy is a constant;

ii-3) Introducing (11.25) and (11.26) into (11.40), under condition (11.42), one obtains the uncertainty relation as an equality ([38], 40])

$$(\delta p)^2 (\delta q)^2 = \frac{1}{16\pi^2 E^2(a, b, c)} + cov^2(p, q) \qquad (11.43)$$

or, as an inequality

$$(\delta p)(\delta q) \geq \frac{1}{4\pi E(a, b, c)}. \qquad (11.44)$$

11.5. Onicescu informational energy of the harmonic oscillator

In the plane (p, q), the energy of a harmonic oscillator of mass m and pulsation ω

$$E = \frac{p^2}{2m} + \frac{m\omega^2 q^2}{2} = \varepsilon_0 = const. \qquad (11.45)$$

represents an ellipse of semi-axes

$$a_0 = \sqrt{2mE}; \qquad b_0 = \sqrt{\frac{2E}{m\omega^2}}. \qquad (11.46)$$

Since $H(p, q)$ appearing in Gaussian (11.23) has to be a dimensionless quantity, to determine the set of parameters (a, b, c) it is necessary the identification

$$H(p,q) \equiv \frac{\varepsilon_0}{E},$$ (11.47)

with

$$\bar{p} = 0; \qquad \bar{q} = 0.$$ (11.48)

We then have

$$a = \frac{1}{mE}; \qquad b = 0; \qquad c = \frac{m\omega^2}{E}.$$ (11.49)

Under these conditions, the Onicescu informational energy of the harmonic oscillator becomes

$$E(a,b,c) = \frac{1}{2}\left(\frac{\nu}{E}\right),$$ (11.50)

where ν is oscillator's frequency. It can be easily verified that (11.41) represents area of the ellipse with semi-axes (11.47)

$$A = \pi a_0 b_0 = \frac{E}{\nu}.$$ (11.51)

Assuming the principle of stationary informational energy (11.42), together with quantum postulate

$$\frac{E}{\nu} = nh; \quad n = 1, 2, 3, \ldots,$$ (11.52)

we can affirm that:

i) The informational energy of the linear harmonic oscillator

$$E(a,b,c) = \frac{1}{2nh}$$ (11.53)

is quantized;

ii) For $n = 1$, (11.53) writes

$$h = \frac{1}{2E(a,b,c)},$$ (11.54)

which allows us to redefine the Planck constant h — a minimum area in the phase space (p,q) — as a maximum value of the informational energy in the phase space (irrespective of the factor 1/2).

iii) Relations (11.43) and (11.53) lead to the minimal form of Heisenberg's uncertainty relation

$$(\delta p)(\delta q) = n\hbar. \tag{11.55}$$

For $n = 1$, (11.53) reduces to

$$(\delta p)(\delta q) = \hbar. \tag{11.56}$$

Relations (11.55) and (11.56) are minimal, because the considered harmonic oscillator is isolated.

11.6. Onicescu informational energy of a system of coupled oscillators

In the one-dimensional case, the differential equation of the harmonic oscillator writes

$$\ddot{q} + \omega^2 q = 0, \tag{11.57}$$

where q is the coordinate. The general solution of this equation is

$$q(t) = h e^{i(\omega t+\phi)} = \overline{h} e^{-i(\omega t+\phi)},$$

where h is a complex amplitude, $\overline{h}$ its complex conjugate, ϕ a specific phase, and t the time. Thus, quantities h and ϕ label each oscillator out of an assembly with the same equation of motion (11.57) and, consequently, the same pulsation ω. We have to do with an assembly of 'Planck resonators' interacting with electromagnetic radiation, whose analysis lead to the law of distribution of energy density in terms of frequency. Such an assembly can be described by a continuous group in three variables with three parameters, as follows [41].

As it is known, the ratio of two fundamental solutions of equation (11.57), denoted as

$$k = e^{2i(\omega t+\phi)},$$

is a solution of Schwartz equation [42]

$$\{k, t\} = 2\omega^2, \tag{11.58}$$

where

$$\{k, t\} = \left(\frac{k''}{k'}\right)' - \frac{1}{2}\left(\frac{k''}{k'}\right)^2$$

is the Schwarzian of function k with respect to t. Equation (11.58) has the property of being invariant with respect to a homographic function transformation, which means that function

$$\sigma(t) = \frac{\alpha k + \beta}{\gamma k + \delta}; \qquad \alpha, \beta, \gamma, \delta \in \Re \tag{11.59}$$

shall satisfy the same equation.

Transformations (11.59) form a continuous group with three real parameters on the complex straight line; this fact confers to parameter k a projective character.

It then follows that in a system of coupled oscillators, each oscillator can be specified by the projective parameter

$$\sigma(t) = \frac{h + \overline{h}k}{1 + k}, \tag{11.60}$$

The oscillators 'synchronization', *i.e.* their correlation in both amplitude and phase, is performed by homographic relation

$$\sigma'(t) = \frac{\alpha\sigma(t) + \beta}{\gamma\sigma(t) + \delta}; \qquad \alpha, \beta, \gamma, \delta \in R. \tag{11.61}$$

Relation (11.61) leads to Barbilian group [43]

$$h' = \frac{\alpha h + \beta}{\gamma h + \delta}; \qquad k' = \frac{\alpha\overline{h} + \beta}{\gamma h + \delta}k \tag{11.62}$$

with Lie vectors

$$B_1 = \frac{\partial}{\partial h} + \frac{\partial}{\partial \overline{h}}; \qquad B_2 = h\frac{\partial}{\partial h} + \overline{h}\frac{\partial}{\partial \overline{h}};$$

$$B_3 = h^2\frac{\partial}{\partial \overline{h}} + \overline{h}^2\frac{\partial}{\partial \overline{h}} + (h - \overline{h})k\frac{\partial}{\partial k}, \tag{11.63}$$

and algebra

$$[B_1, B_2] = B_1; \qquad [B_2, B_3] = B_3; \qquad [B_3, B_1] = -2B_2. \qquad (11.64)$$

Since groups (11.1) and (11.62) are isomorphic, a theorem which belongs to Stoka [44] allows us to construct the functions which are simultaneously invariant under the action of groups (11.1) and (11.62), as solutions of equations

$$L_\ell F(x, y, h, \overline{h}, k) + B_\ell F(x, y, h, \overline{h}, k) = 0; \quad \ell = \overline{1,3} \qquad (11.65)$$

which are the functions $H(x, y)$. It follows that

$$F(\mu, \nu) = const. \qquad (11.66)$$

with

$$\mu = \frac{-i(h - \overline{h})}{(x - hy)(x - \overline{h}y)};$$
$$\nu = \frac{x - \overline{h}y}{x - hy}k. \qquad (11.67)$$

A particular class of such invariant functions is represented by the linear combinations

$$p\mu = m\left(\nu + \frac{1}{\nu}\right) + 2n, \qquad (11.68)$$

where p, m, n are arbitrary constants. In view of (11.67), relation (11.67) can also be written as

$$mk^{-1}z^2 + 2nz\overline{z} + mk\overline{z}^2 = p, \qquad (11.69)$$

with

$$z = \frac{x - \overline{h}y}{\sqrt{-i(h - \overline{h})}}, \qquad (11.70)$$

together with condition

$$i(h - \overline{h}) > 0. \qquad (11.71)$$

Equation (11.69) represents a family of conics situated in the plane (x, y). They are ellipses if

$$m^2 - n^2 < 0, \tag{11.72}$$

This condition is satisfied if

$$m = Q \sinh(2u);$$
$$n = Q \cosh(2u), \tag{11.73}$$

where Q is a real constant. Now, it is allowed to take h as a purely imaginary number, $h = i$, without restraining generality. In this case the quadratic form (11.69), identified with $H(x, y)$ given by (11.24), produces the following expressions for (a, b, c):

$$a = Q[\cosh(2u) + \sinh(2u) \cos \phi];$$
$$b = -Q \sinh(2u) \sin \phi; \tag{11.74}$$
$$c = Q[\cosh(2u) - \sinh(2u) \cos \phi],$$

where ϕ is the value of k, supposed to be fixed. The Onicescu informational energy becomes

$$E(a, b, c) = \frac{Q}{4\pi}, \tag{11.75}$$

so that the uncertainty relation (11.43) takes the form

$$(\delta p)^2 (\delta q)^2 = \frac{1}{Q^2}[1 + \sinh^2(2u) \sin^2 \phi]. \tag{11.76}$$

If

$$Q = \frac{2}{\hbar}, \tag{11.77}$$

formula (11.76) reduces to the uncertainty relation given by Stoler [39] for coherent states

$$(\delta p)^2 (\delta q)^2 = \frac{\hbar^2}{4}[1 + \sinh^2(2u) \sin^2 \phi]. \tag{11.78}$$

The uncertainty has a minimum only if $\phi \equiv 0$, meaning that all oscillators of the assembly have the same initial phase, which is zero.

Under this hypothesis, at any subsequent moment of time the uncertainty relation (11.78) leaves its minimality condition, together with covariance of assembly which becomes different from zero.

11.7. Observable quantities and Onicescu informational energy

Since a measurement is an act of interaction given by the correspondence

$$(p, q) \to (p', q'), \tag{11.79}$$

the following question arises: which is the set of quantities determined by measurement, (p, q) or $(p'q')$? To answer this question, let us suppose that measurement is characterized by transformation [39]

$$q' = aq + bp + c;$$
$$p' = cq + dp + e. \tag{11.80}$$

If the result of measurement is $(\overline{p}, \overline{q})$, then it is necessary to have

$$\overline{q} = a\overline{q} + b\overline{p} + c;$$
$$\overline{p} = c\overline{q} + d\overline{p} + e, \tag{11.81}$$

which shows that two parameters of the transformation can be explained, so that the number of independent parameters reduces to four. Extracting c, e from (11.81) and introducing their values into (11.80), we are left with

$$q' - \overline{q} = a(q - \overline{q}) + b(p - \overline{p});$$
$$p' - \overline{p} = c(q - \overline{q}) + d(p - \overline{p}). \tag{11.82}$$

On the other hand, according to [39], the group (11.82) is measurable if it is unimodular. It follows that it shall depend on three essential parameters, this way being justified not only our preference for Gaussians of the same mean value, but also the whole previous formalism. This means that the measurable — or observable — quantities (p, q) are mean values $(\overline{p}, \overline{q})$ of the Gaussian distributed variables, associated with a constant informational energy.

The fact that the group is measurable allows one certain correspondences with quantum mechanics. Let us exemplify this property for the Barbilian group (11.62), whose elementary measure is

$$d\Omega = \frac{dh \wedge d\overline{h} \wedge dk}{(h - \overline{h})^2 k}.$$ (11.83)

Then the invariantive function of the group

$$\Psi = -\frac{1}{(h - \overline{h})^2 k} = \frac{1}{v} e^{\frac{i}{2}\varphi},$$ (11.84)

where $h = u + iv$ and $k = e^{i\varphi}$, may have significance of "wave function" of a stationary Schrödinger equation

$$\nabla^2\Psi = \left[\left(\frac{\nabla v}{v}\right)^2 + \frac{\nabla u \nabla \varphi}{v} - \frac{(\nabla\varphi)^2}{4} + \frac{i}{2}\frac{\nabla u \nabla \varphi}{v} \right] \Psi,$$ (11.85)

with a complex "proper value". This becomes real if either the oscillators have the same phase, or the same amplitude. The first situation is very interesting, because equation (11.85) reduces to

$$\nabla^2\Psi = \left(\frac{\nabla u \nabla \varphi}{v}\right)^2 \Psi$$ (11.86)

with squared momentum as "proper value", which is responsible for momentum-coordinate uncertainties of hydrodynamic model of quantum mechanics. Therefore, we can conclude that the uncertainty relations (11.43) and (11.44) can also be interpreted in case of theories with hidden parameters.

11.8. Onicescu informational energy of the free particle

According to invariantive mechanics, energy of the free particle is

$$E = \varepsilon m\omega^2; \qquad \varepsilon = \pm 1,$$

where m and ω are the mass and velocity limit of the particle, respectively. If we consider the energy mean value, deduced under conditions of both minimum informational entropy and Steinhaus' theorem [37], the probability density of velocity is

$$\rho = e^{-\frac{mv^2}{2E}} = e^{-\frac{v^2}{2\varepsilon\omega^2}}. \tag{11.87}$$

Under these conditions, the corresponding Onicescu informational energy

$$E(\omega) = \int_{-\infty}^{+\infty} e^{-\frac{v^2}{2\omega^2}}\, dv \tag{11.88}$$

with restriction

$$\varepsilon\omega^2 > 0, \tag{11.89}$$

becomes

$$E(\omega) = k\omega, \tag{11.90}$$

where k is a constant. It then follows that:

i) Existence of Onicescu informational energy of a free particle excludes situation $\varepsilon = -1$;

ii) Acceptance of stationarity principle of Onicescu informational energy of a free particle comes back to condition

$$\frac{H}{m} = \omega^2 = const. \tag{11.91}$$

which is equivalent to one of the principles of invariantive mechanics;

iii) If $\omega = c$, that is in case of special relativity, the speed of light in vacuum shall be redefined — up to an arbitrary factor $1/k$ — as Onicescu informational energy of a free particle.

11.9. Informational energy during the process of interaction neutron — crystal net

As it is known, the atom motion in crystal net can be described by means of coherent states ([39], [41]) which are proper states of the annihilation operator of the bosonic field (the coherent states are detailed in the Appendix to [39]). This method of description of atomic motion is connected with an essential advantage: the thermodynamic state sums can be easily constructed by means of certain density operators. These operators characterize the assemblies of harmonic motions and, in their turn, are constructed by means of the maximum informational entropy principle. This procedure is described in [39] for the quantum assembly characterizing the simultaneous determinations of canonically conjugated quantum variables position — momentum. The result for the density operator characterizing the state immediately after measurement [39] is:

$$\hat{\rho}_m = \frac{\hbar}{\Delta p \Delta q + \frac{\hbar}{2}} \exp\left\{ -\frac{\hbar}{\Delta p \Delta q + \frac{\hbar}{2}} (\hat{a}^+ - \overline{\alpha})(\hat{a} - \alpha) \right\} \qquad (11.92)$$

with

$$\hat{a} = \frac{1}{\sqrt{2\hbar\mu}} (\hat{a} + i\mu\hat{p}). \qquad (11.93)$$

Here $\hat{p}$ and $\hat{q}$ are momentum and coordinate operators, respectively, p and q being their mean values. The complex quantity

$$\alpha = \frac{1}{\sqrt{2\hbar\mu}} (q + i\mu p) \qquad (11.94)$$

is the proper value of the annihilation operator (11.93) on coherent states

$$\hat{\alpha}|\alpha\rangle = \alpha|\alpha\rangle. \qquad (11.95)$$

We mention that reasoning which led to the operator (11.92) presuppose an aprioric explicitation of variations Δq and Δp, involved in the notation

$$\mu = \frac{\Delta q}{\Delta p}. \qquad (11.96)$$

This fact is natural, if we want to apply the principle of maximum of informational entropy. We nevertheless insist on its significance to draw attention on the fact that the operator (11.93) becomes an oscillator characteristic for the particular value

$$\mu_0 = (M\omega)^{-1}, \tag{11.97}$$

where M and ω are the mass and pulsation of the oscillating particle. Operator (11.93) then takes the already known form

$$\hat{a} = \sqrt{\frac{M\omega}{2\hbar}}\left(\hat{q}_0 + i\frac{\hat{p}_0}{M\omega}\right). \tag{11.98}$$

When measurement satisfies the minimum "uncertainties", that is

$$\Delta p \Delta q = \frac{\hbar}{2}, \tag{11.99}$$

$\hat{\rho}_m$ occurring in (11.92) is projection on proper state $|\alpha >$ of the annihilation operator

$$\hat{\rho}_m = |\alpha\rangle\langle\alpha|. \tag{11.100}$$

By means of the above displayed formalism, let us now describe the interaction between a neutron and an atom of the crystal net, assimilating the elastic collision with a process of simultaneous measurement of the coordinate and momentum of the atom. The state immediately after collision shall be described by a density operator, the mean value of the atom energy obtained as a result of action of the operator being the energy of the primary collided atom. In the following analysis we shall consider only the coordinate and momentum magnitudes, on spherical symmetry grounds to make the as clear as possible.

In an elastic collision, if p_0 is the initial momentum of the projectile particle, and P_0 that of the target particle, the momentum of the target after collision is [39]

$$\mathbf{p} = -MV\mathbf{n} + \frac{A}{1+A}(\mathbf{P}_0 + \mathbf{p}_0); \qquad M = \frac{Mm}{M+m}, \tag{11.101}$$

where M is the reduced mass of the system, m — the mass of projectile particle, and A — the ratio between masses of target and projectile. In the concrete case of atom-neutron, A stands for the mass number of the element the target is made out of. Some elements in formula (11.101) are determined up to a certain degree of precision. All we can specify here are the relative velocity V and the initial momentum $\mathbf{p}_0$, even if the last quantity has a certain degree of indecision. There are some elements that cannot be specified in (11.101), such as direction of target after collision, determined by versor $\mathbf{n}$, and the target momentum $\mathbf{P}_0$ before collision, which is a result of action of collective oscillations at a certain moment. In this category also falls the coordinate $\mathbf{q}_0$; to use a one-dimensional theory, its direction can be associated with $\mathbf{P}_0$. In addition, there exists an 'assembly' indecision regarding direction of $\mathbf{p}_0$: in a chaotic flux of projectile-particles, the target can be a priory collided from any direction. Finally, there is an indecision related to the fact that we do not effectively know the place of collision, around the equilibrium position of the target. All these indecisions leave their specific mark on the data of our problem.

First, we are obliged to fully ignore directions of $\mathbf{n}$ and $\mathbf{p}_0$ while discussing their distribution on the sphere of unit radius with natural density

$$\rho(\theta, \varphi) = \frac{1}{4\pi} \sin \theta \, d\theta \, d\varphi, \tag{11.102}$$

where $\theta \in [0, \pi]$, $\varphi \in [0, 2\pi)$. The averaging of (11.101) over the directions assembly furnished by (11.102) leads to the vector

$$\mathbf{p} = \frac{A}{1 + A} \mathbf{P}_0, \tag{11.103}$$

with equal variants on its components, given by

$$\Delta p = \frac{A}{\sqrt{2}(1 + A)} p_0. \tag{11.104}$$

In order to approach indecision regarding the place of collision, we have to appeal to a quantity which belongs to nuclear interactions field, namely the effective cross section. As it is well known, both elastic and inelastic collisions, considered as reactions between particles, can be characterized by the reaction effective cross section, which is a measure of the probability of process appearance. Within elastic limits, this can be considered as a measure of the target area exposed to the projectile fascicle. In this respect, we agree upon hypothesis that both target and projectile have to be mutually situated in a vicinity of linear dimensions given by the square root of the process effective section. This quantity shall be taken as variance of the target coordinate. Therefore, we shall have

$$\Delta q = \frac{1}{\sqrt{3}}\sqrt{\sigma}, \qquad (11.105)$$

where σ is the effective section, while Δq is the same for all three degrees of freedom. Using the commutation relations for the quantum operators associated with coordinate and momentum, we have

$$[\hat{q}_m, \hat{p}_n] = i\hbar\delta_{MN}; \qquad m, n = \overline{1,3}. \qquad (11.106)$$

This imposes the choice of the radius vector immediately after collision as

$$\mathbf{q} = \frac{A+1}{A}\mathbf{q}_0, \qquad (11.107)$$

since we consider q_0 and p_0 as being canonically conjugated operators.

Making an inventory of our discussion, the state of the target immediately after collision shall be characterized by the operator (11.92), corresponding to one degree of freedom, where α from (11.94) is given by

$$\alpha = \sqrt{\frac{A\sqrt{\sigma}}{2\hbar(A+1)\sqrt{2mE}}}\left(\frac{A+1}{A}q_0 + i\sqrt{\frac{\sigma}{2mE}}p_0\right). \qquad (11.108)$$

Here E is the kinetic energy of the projectile, and

$$p_0 = \sqrt{2mE}. \qquad (11.109)$$

The collision problem is now posed as follows: the target (atom) is brought from the state characterized by (11.92), with α_0 given by (11.98), to the state defined by (11.92) and α given by (11.108). Certainly, one can define an operator of Bogoliubov-type [4] associated with α expressed by (11.108), such as

$$\hat{a} = \hat{U}_r \hat{\alpha}_0 \hat{U}_r^+. \tag{11.110}$$

Here $\hat{U}_r$ is the unitary operator

$$\hat{U}_r = \exp\left\{\frac{1}{2}r\left(\hat{a}^2 - (\hat{a}^+)^2\right)\right\}, \tag{11.111}$$

and we denoted

$$r = \ln\frac{A}{A+1}\sqrt{\frac{\mu_0}{\mu}}, \tag{11.112}$$

with μ_0 given by (11.97). A new notation

$$\nu = \frac{\hbar}{\Delta p \Delta q + \frac{\hbar}{2}}, \tag{11.113}$$

together with (11.111), give rice to the following form of operator (11.92)

$$\hat{\rho}_m = \nu\hat{U}_r \exp\left\{-\nu\langle \hat{a}_0^+ - \overline{\alpha}|\hat{a}_0 - \alpha\rangle\right\}\hat{U}_r^+, \tag{11.114}$$

where between α and α_0 (the proper value of $\hat{a}_0$) there exists the following $SL(2R)$ transformation

$$\alpha = \cosh r\alpha_0 + \sinh r\overline{\alpha}_0. \tag{11.115}$$

After all these considerations, we are asked to take into account a certain type of indecision, connected to the assembly of collective oscillations existent in crystal net: the target (atom) is characterized by the operator $\hat{a}_0$, which is specific to an oscillator belonging to a thermic assembly, characterized by the density operator

$$\hat{\rho}_T = (e^{\beta\hbar\omega} - 1)\int \frac{d^2z}{\pi} exp\{1 - e^{\beta\hbar\omega}|z^2|\}|z\rangle\langle z|, \tag{11.116}$$

where T is the absolute temperature, $\beta = 1/kT$, and k is Boltzmann's constant. Consequently, it is necessary to perform averaging of (11.114) over the thermic assembly (11.116). As one knows [39], this operation reduces to calculation of the trace of the operators product, meaning an integral in the complex plane z with the usual measure

$$\frac{d^2 z}{2\pi} = \frac{dq\, dp}{2\pi},$$

leading to

$$TR(\hat{\rho}_m \hat{\rho}_r) = \frac{\nu}{1 + \overline{n}\nu} \exp\left\{ -\frac{\nu |\alpha|^2}{1 + \overline{n}\nu} \right\}, \tag{11.117}$$

where ν is given by (11.113) and we denoted

$$\overline{n} = (e^{\beta \hbar \omega} - 1)^{-1}. \tag{11.118}$$

Therefore, the state of the target immediately after collision, at the temperature of the crystal net, is given by the operator

$$\hat{\rho} = \frac{\nu}{1 + \overline{n}\nu} \exp\left\{ -\frac{\nu}{1 + \overline{n}\nu} \hat{a}^+ \hat{a} \right\}. \tag{11.119}$$

The density operator — or, better, the repartition density (11.117) — stands for a typical Gaussian assembly. The informational energy is then given by

$$TR\left(\hat{\rho}^2\right) = 2\frac{1 + \overline{n}\nu}{\nu} - 1, \tag{11.120}$$

which is a natural generalization of classical formula of this statistical quantity.

In view of (11.119), the principle of stationary informational energy receives a new interpretation: the mean energy of the assembly is a constant. Indeed, the Hamiltonian operator is given in the usual quantum form

$$\hat{H} = \frac{1}{2}(\hat{a}^+ \hat{a} + \hat{a}\hat{a}^+), \tag{11.121}$$

while the mean value of this quantity, given by

$$TR\left(\hat{\rho}\hat{H}\right) = \overline{H}, \tag{11.122}$$

as a result of a simple calculation which takes into account the density (11.117), allows us to write

$$\hat{H} = \frac{1}{2} + \frac{1 + \overline{n}\nu}{\nu}. \tag{11.123}$$

Since this quantity differs only by a constant from the informational energy, the above statement is justified. Formally, $\hat{H}$ is identical to parenthesis of

$$\rho(\omega, T) = \frac{\hbar\omega^3}{2\pi^2 c^3}\left(\frac{1}{2} + \frac{1}{e^{\frac{\hbar\omega}{kT}} - 1}\right),$$

but never reduces to it. Anyhow, $\hat{H}$ simultaneously contains both statistics and thermodynamics.

As a matter of fact, the quantity $\hat{H}$ of (11.123) is an abstract number. What is really determined is $\hat{H}$ multiplied by an energetic constant, and has to be the analogous of the coefficient met in electromagnetic interactions given by $\rho(\omega, T)$. Since such an investigation — connecting electromagnetic interactions with some other types of interactions — does not exist, we shall adopt the value of a "quantum" for the energetic constant

$$\varepsilon_0 = \frac{\hbar}{M\mu} = \frac{\hbar}{A + 1}\sqrt{\frac{2E}{m\sigma}}, \tag{11.124}$$

where E is the energy of the projectile particle, and m its mass. Here we keep in mind the idea that the fundamental interaction is electromagnetic, but this time it is not produced by the photon. If we adopt (11.97) as value for μ, formula (11.124) goes to the usual one for an oscillator; for the moment, this is its only viable verification. Another justification is, for example, calculation of the energy of a collided primary atom and its comparison with the unanimously accepted data, which are not quite sure from the determinations point of view [39]. Nevertheless, we are satisfied to observe that our formula, together with its deduction, are pretty close to reality.

Indeed, using (11.123) and (11.124), we obtain

$$\overline{H}\varepsilon_0 = \frac{2A}{3(A+1)^2}E + \frac{\hbar}{A+1}\sqrt{\frac{2E}{m\sigma}}\left(\coth\frac{\beta\hbar\omega}{2} + 1\right). \qquad (11.125)$$

Nowadays, the neutrons producing defects in crystal net are typically situated in the rapid energy specter, with values of 1 MeV or higher. The mean value for effective section of elastic collision for zirconium — as an example — decreases very rapidly for energies of $2 - 3$ MeV. If, for the same element, we consider a typical frequency of phononic specter which is about 1.5×10^{19} c.p.s., formula (11.125) offers the value of 2.26×10^4 eV for the energy of collided primary atom. The estimated values used for practical purposes take into account only half of the maximum of the collided atom (4.3×10^4 eV), that is $2.15 \times 10^4 eV$, which is in a very good agreement with our theoretic calculation. Since such a quantity can be determined only by indirect measurements, it is affected by a high level of arbitrarity. Formula (11.125) can be successfully verified on a large set of mono-atomic materials. It is more than necessary to calculate this way certain quantities of high interest; it can also be useful in the study of interactions between particles and solid state matter.

11.10. Onicescu informational energy and invariantive mechanics

Admitting the principle of stationarity of Onicescu informational energy in invariantive mechanics, analysis of the results obtained in this paragraph lead to the following conclusions:

i) Certain postulates of invariantive mechanics are eliminated, such as:

- postulate $\frac{H}{m} = const.$ in case of inertial motion of material point;
- postulate regarding cancellation of the exterior product of two 1-forms in case of de Broglie linear theory;

- postulate concerning equality of external differentials of two 1-forms, in case of the motion of a particle in a linear field;
- postulate regarding equality of exterior differential of 1-form with the field 2-form in case of motion of a particle in an arbitrary field, etc.

ii) Certain fundamental physical constants, such as Planck's constant $\hbar$ and speed of light c are introduced and, also, redefined.

iii) All 1-forms and 2-forms corresponding to a fundamental physical constant are submitted to the same class of statistical hypotheses.

iv) The alternative $\varepsilon = -1$ is excluded in invariantive mechanics, but the alternative of negative limited velocities is accepted. In this last case, the invariantive model and the anisotropic relativity are equivalent [39].

References (Chapters 1–11)

[1] M. Agop, C. Dariescu, M. Dariescu, *Lés équations Lagrange, Hamilton et Hamilton-Javobni sur la variété $E_3 \times T$*. Bulletin of the Polytechnic Institute of Iasi, Tom XXXVIII(XLII), Fasc. 1-4, Mathematics, Theoretical Mechanics, Physics, 1992, pp. 193–195.

[2] I. I. Placinteanu, *Vectorial and Analytical Mechanics* (in Romanian), Technical Publishing House, Bucharest, 1958, 2nd Ed.

[3] O. Onicescu, *Mechanics* (in Romanian), Technical Publishing House, Bucharest, 1959.

[4] I. Merches, D. Tatomir, R. E. Lupu, *Basics of Quantum Electrodynamics*, CRC Press, 2012.

[5] O. Onicescu, *Invariantive Mechanics*, Springer-Verlag, Wien, 1975.

[6] M. Agop, G. Ciobanu, P. D. Ionnou, C. Buzea, *Cantorian $\varepsilon^{(\infty)}$ Space-Time in Cartan, de Boglie, and Fields Theories*, Chaos, Solitons and Fractals, Vol.14, 2002, pp. 863–890.

[7] M. Agop, C. Craciun, *Cartan Gravitation in Fractal Space-Time*, Particles and Fields, Editors: M. Agop, P. D. Ioannou, 2005, Athens University Press, pp. 75–104.

[8] J. Argyris, C. Ciubotariu, H. G. Matuttis, *Fractal Space. Cosmic Strings and Spontaneous Symmetry Breaking*, Chaos, Solitons and Fractals, Vol. 12, 2001, pp. 1–48.

[9] M. Agop, H. Matsuzawa, I. Oprea, R. Vlad, C. Sandu, C. Buzea, *Some Implications of the Gravito-magnetic Field in Fractal Space-Time Theory*, Australian Journal of Physics, 53, 2000, pp. 217–230.

[10] N. Mazilu, M. Agop, *A Particle Model in the Invariantive Mechanics*, Scientific Annals of the Al. I. Cuza University of Iasi, Tom XXIX, New Series, Section 1, Physics, 1983, pp. 56–61.

[11] M. Agop, N. Mazilu, Contributions to Development of Invariantive Mechanics (in Romanian), Scientific Sections Memories, Series IV, Tom VIII, nr.1, 1985, pp. 113–132.

[12] M. Agop, M. Dariescu, C. Dariescu, *Invariantive Hydrodynamics and Electrodynamics* (in Romanian), Scientific Sections Memories, Series IV, Tom XII, nr.1, 1989, pp. 131–144.

[13] M. Agop, A. Stroe, *Linear and Nonlinear Cartan Fields Theories in Fractal Space-Time*, Particles and Fields, Editors: M. Agop, P. D. Ioannou, 2005, Athens University Press, pp. 105–119.

[14] E. Cartan, *Les systèmes différentiels extérieurs et leurs applications géométriques*, Actualités Sci. Ind., no. 994, Hermann et cie., Paris, 1945.

[15] H. Flanders, *Differential Forms with Applications for the Physical Sciences*, London Academic Press, New York, 1963.

[16] J. Argyris, C. Marin, C. Ciubotariu, *Physics of Gravitation and the Universe*, Vol. II, Fractal and Cosmological Large-Scale Structure, Ed. Spiru Haret, Iasi 2012, Ed. Tech.-Info, Chisinau, 2002.

[17] Corina Marin, *Contributions to the Study of the Linear Theory of the Gravitational Field* (PhD Thesis — in Romanian), Iasi, 2002.

[18] L. D. Landau, E. M. Lifshitz, *Electrodynamics of Continuous Media* (Volume 8 of A Course of Theoretical Physics), Pergamon Press, 1960.

[19] F. Halbwachs, *Théorie relativiste des fluids a spin*, Gauthier-Villars, Paris, 1960.

[20] G. t Hooft, Nuclear Physics B, vol. 190, 1981, pp. 455–478.

[21] L. de Broglie, *Mcanique ondulatoire du photon et théorie quantique des champs.* Gaurhier-Villars, Paris, 1957.

[22] L. de Broglie, *La Thermodynamique de la particule isolée (ou Thermodynamique cachée des particules)*, Gauthier-Villars, Paris, 1964.

[23] I. Mihaila, C. R. Acad. Sci. Paris 280, Serie A, 1975, pp. 595.

[24] I. Mihaila, Rev. Roum. Phys. 25, 1980, p. 3.

[25] J. Chazy, *La Théorie de la Relativité et la Mécanique Céleste*, T. I. Chap. 3, Gauthier-Villars, Paris, 1928.

[26] I. Mihaila, *The Motion of Perihelion* (in Romanian), Note in Invariantive Mechanics and Cosmology by O.Onicescu, Academy Publishing House, Bucharest, 1974, pp. 96–100.

[27] M. Agop, *Contributions à l'étude de l'interaction gravitationelle dans la mécanique invariantive*, Rev. Roum. de Physique, Tome 27, Nr. 3, 1982, pp. 219–234.

[28] M. Agop, *La déviation de la lumière dans la mécanique invariantive*, Rev. Roum. de Physique, Tome 25, Nr. 3, 1980, pp. 295–299.

[29] V. A. Fock, *The Theory of Space, Time and Gravitation*, Macmillan, 1964.

[30] N. Ionescu-Pallas, *General Relativity and Cosmology* (in Romanian), Scientific and Encyclopedic Publishing House, Bucharest, 1980.

[31] L. Sofonea, *Invariance Principles of the Theory of Motion* (in Romanian), Academy Publishing House, Bucharest, 1973.

[32] M. Agop, D. Tatomir, C. Buzea, Cristina Buzea, A. Zacharias, Journal of Franklin Institute, vol. 334B, No. 1, 1997, pp. 57–62.

[33] M. Agop, *Les lois de la variation de la mass avec le temp. La corrélation mass-distance.* Revue Roumaine des Sciences Technique, Serie de Mécanique Appliqée, Tome 33, Nr. 1, 1988, pp. 33–37.

[34] M. Agop, S. Talasman, Invariantive theory of the rigid body (in Romanian), Memoriile Sectiilor Stiintifice, Seria IV, Tomul XI, br.1, 1988, pp. 107–132.

[35] F. Bowman, *Introduction to elliptic functions with applications*, London, English University Press, 1955.

[36] I. Merches, M. Agop, *Remarks on the relativistic invariantive magnetofluid dynamics*, Bulletin of the Polytechnic Institute of Iasi, Tom XXV(XXIX), Fasc. 3–4, Section I, Mathematics, Theoretical Mechanics, Physics, 1979, pp. 93–96.

[37] O. Onicescu, *Probability and Random Processes* (in Romanian), Scientific and Encyclopedic Publishing House, Bucharest, 1977.

[38] M. Agop, Cristina Buzea, C. Buzea, L. Chirila, S. Oancea, *On the Information and Uncertainty Relation of Canonical Quantum System with SL(2R) Invariance*, Chaos, Solitons and Fractals, vol. 7, No. 5, 1996, pp. 659–668.

[39] N. Mazilu, M. Agop, *Physics of the Measurement process* (in Romanian), Stefan Procopiu Publishing House, Iasi, 1994.

[40] M. Agop, V. Melnig, *L'énergie informationnelle et les rélations d'incertitude pour les systèmes canoniques SL(2R) invariants*, Entropie, No. 188/189, 1995, pp. 119–123.

[41] M. Agop, V. Griga, B. Ciobanu, Cristina Buzea, C. Stan, D. Tatomir, *The Uncertainty Relation for an Assembly of Plank-type Oscillators. A Possible GR-Quantum Mechanics Connection*, Chaos, Solitons and Fractals, vol. 8, No. 5, 1997, pp. 809–821.

[42] M. Mihaileanu, *Differential, Projective, and Analytical Geometry* (in Romanian), Didactic and Pedagogical Publishing House, Bucharest, 1972.

[43] D. Barbilian, *Didactic opera; elementary geometry* (in Romanian), Technical Publishing House, Bucharest, 1968.

[44] M. Stoka, *Integral Geometry* (in Romanian), Academy Publishing House, Bucharest, 1968.

Chapter 12

Chaos Via Fractality
in Gravitational Dynamical Systems

12.1. Introduction

As well known, the chaotic type phenomena are relevant in a large variety of science fields. Classical theory of physics defines the chaotic behavior as an extreme sensitivity (hypersensitivity) to variation in the initial conditions of the equation describing the system dynamics. Subsequently, the systems that exhibit mathematical chaos are deterministic and thus ordered in some sense. Such nonlinear dynamical systems are, for example, the solar system, the terrestrial atmosphere, turbulent fluids, the electronic generators and circuits ([1]–[4]).

However, in many systems where chaos arises, spatial and temporal structures seem also to appear, being frequently experimentally observed in chemistry (Belousov-Zhabotinsky reaction), electronics (Chua-Matsumoto circuit), fluid dynamics (Rayleigh-Bernard convention, Taylor-Couette flow) etc. These structures are in some few cases determined or confirmed by means of numerical simulations, but very rarely understood or predicted from a fundamental theory [5].

The gravitational structures make no exception from the above mentioned law. One of the most impressive examples is the fruitful suggestion made by Laskar that the inner planetary system (telluric planets) is chaotic (Laskar, [6])]. It was also experimentally observed the regularity of the distribution of planets, satellites and asteroids in the solar system ([7], [8]).

In the following analysis we propose a general method for approaching both chaos and pattern formation by presenting the nonlinear interaction between a neutral particle beam and a composite field (a gravito-electromagnetic field overlapping a constant external gravito-magnetic field). Such a physical situation frequently appears in gravitational binary star systems ([9], [10]).

This chapter is structured in six sections. In paragraph 2, the gravitational Maxwell-type equations are presented. Further on, in paragraph 3, the equations of motion and their numerical solutions are obtained. The Lyapunov exponent analysis, bifurcation diagrams and fractal analysis are performed in sections 4, 5, and 6, respectively.

12.2. Gravitational Maxwell-type equations

Consider the four-dimensional space-time as being split into space plus time $(3+1)$. In this case, the general relativistic gravitational field (the contravariant form of the metric tensor) breaks into three parts ([11], [13]):

1. An electric-like part, g^{00}, whose gradient for weak gravity is the Newtonian acceleration $\mathbf{g}_g$;
2. A magnetic part, g^{0i}, whose curl, for weak gravity, is the gravito-magnetic field $\mathbf{B}_g$;
3. A spatial metric, g^{ij}, whose curvature tensor is the "curvature of space".

The relativistic weak-field approximation starts with the postulate that the local coordinates, covering some patch on the space-time manifold, form an open neighborhood about the origin, in terms of which the metric is Minkovskian up to first order in coordinate over this patch. This construction may be carried out everywhere on the manifold.

In the weak (gravitational) field approximation, the metric tensor $g_{\mu\nu}$ can be written as

$$g_{\mu\nu} = \eta_{\mu\nu} + h_{\mu\nu} = \eta_{\mu\nu} + \overline{h}_{\mu\nu} - \frac{1}{2}\eta_{\mu\nu}\overline{h}, \qquad (12.1)$$

where $\eta_{\mu\nu} = (+1, -1, -1, -1)$ is the usual Minkowski metric, $h_{\mu\nu}$ is a small perturbation of $\eta_{\mu\nu}$, and

$$h_{\mu\nu}\overline{h}_{\mu\nu} - \frac{1}{2}\eta_{\mu\nu}\overline{h} \; ; \quad (\mu, \nu = 0, 1, 2, 3). \tag{12.2}$$

Within the same approximation, the contravariant form of the metric tensor writes

$$g^{\mu\nu} = \eta^{\mu\nu} - h^{\mu\nu} \; ; \quad h^{\mu\nu} = \eta^{\mu\alpha}\eta^{\nu\beta}H_{\alpha\beta}, \tag{12.3a, b}$$

while the determinant of $g_{\mu\nu}$ is

$$g = \det(g_{\mu\nu}) \approx -1 - h; \quad (-g)^{1/2} \approx 1 + \frac{h}{2}. \tag{12.4a, b}$$

Since all the covariant derivatives reduce to partial derivatives

$$\nabla_{\mu}h_{\alpha\beta} \approx \partial_{\mu}h_{\alpha\beta}, \tag{12.5}$$

the Riemann tensor $R^{\sigma}{}_{\mu\nu\lambda}$, the Ricci tensor $R_{\mu\nu}$. and the scalar curvature R write:

$$R^{\sigma}{}_{\mu\lambda\mu} \approx \frac{1}{2}\left(\partial^{\sigma}\partial_{\lambda}h_{\mu\nu} + \partial_{\mu}\partial_{\nu}h_{\lambda}^{\sigma} - \partial_{\mu}\partial_{\lambda}h_{\nu}^{\sigma} - \partial_{\nu}\partial^{\sigma}h_{\lambda\mu}\right), \tag{12.6}$$

$$R_{\mu\nu} = R^{\lambda}{}_{\mu\lambda\nu} \approx \frac{1}{2}\left(\partial^{\lambda}\partial_{\lambda}h_{\mu\nu} + \partial_{\mu}\partial_{\nu}h - \partial_{\mu}\partial_{\lambda}h_{\nu}^{\lambda} - \partial_{\nu}\partial^{\lambda}h_{\lambda\mu}\right), \tag{12.7}$$

and

$$R = \eta^{\mu\nu}R_{\mu\nu}, \tag{12.8}$$

respectively.

Using Einstein's equations with the substitutions [12], [13])

$$E^i = G^{00i} \; ; \; B^1 = G^{023} \; ; \; B^2 = G^{031} \; ; \; B^3 = G^{012} \; (i = 1, 2, 3) \tag{12.9}$$

where

$$
\begin{cases}
G^{\alpha\beta\gamma} = \dfrac{1}{4}\left(\overline{h}^{\alpha\beta,\gamma} - \overline{h}^{\alpha\gamma,\beta}\right); \\[2ex]
\mathbf{E} = -\nabla\Phi - \dfrac{\partial\mathbf{A}}{\partial t}; \quad \mathbf{B} = \nabla\times\mathbf{A}; \\[2ex]
\Phi = -\dfrac{1}{4}\overline{h}_{00}; \quad A^i = \dfrac{1}{4}h^{0i},
\end{cases}
\tag{12.10}
$$

and imposing the de Donder gauge conditions

$$
\begin{cases}
\overline{h}^{\alpha\beta}_{,\beta} = \left(h^{\alpha\beta} - \dfrac{1}{2}\eta^{\alpha\beta}h\right)_{,\beta}; \\[2ex]
h = \eta^{\alpha\beta}h_{\alpha\beta} = -\overline{h},
\end{cases}
\tag{12.11}
$$

the Maxwell-type equations for gravito-electromagnetic field are obtained in the usual form

$$
\begin{cases}
\nabla\cdot\mathbf{E} = -\dfrac{\rho}{\varepsilon}; \quad \nabla\cdot\mathbf{B} = 0. \\[2ex]
\nabla\times\mathbf{E} = -\dfrac{\partial\mathbf{B}}{\partial t}; \quad \nabla\times\mathbf{B} = -\mu\mathbf{j} + \varepsilon\mu\dfrac{\partial\mathbf{E}}{\partial t}.
\end{cases}
\tag{12.12}
$$

In the previous relations E^i and B^i define the electric-type and magnetic-type components of the gravitational field (gravito-electric and gravito-magnetic fields, respectively), ε is the gravitational permittivity of vacuum

$$
\varepsilon = \frac{1}{4\pi G},
\tag{12.13}
$$

μ is the gravitational permeability of vacuum

$$
\mu = \frac{4\pi G}{c^2},
\tag{12.14}
$$

j^i are the components of the mass current density, G is Newton's constant, and c is the speed of light in vacuum [14].

This form of the decomposed equations is valid in any coordinate system. They are covariant under transformations of the spatial coordinates $x^i = x^i(x^1, x^2, x^3)$, but it should be stressed that, in general, once a particular coordinate system is chosen, the equations cannot be taken simply over another system of coordinates by an arbitrary coordinate transformation of the type $\bar{x}^\sigma = \bar{x}^\sigma(x^0, x^1, x^2, x^3)$. It is necessary to return to the original term of the decomposed equations and substitute the quantities according to the new system of coordinates.

Quantitatively, the linear approximation is very well justified on the surface of any member of our solar system. Even on the surface of a white dwarf, the deviation of g_{00} from unity does not exceed a value of the order of 10^{-5}. Therefore, we can postulate, *ab initio*, the Maxwell-type gravitational equations for linear gravitation.

12.3. Equations of motion and their numerical solutions

Let us consider the interaction between a beam of particles and a composite field, composed by a gravito-electromagnetic field overlapping a constant external gravito-magnetic field. The mathematical model of such a dynamical system is described by the following equations:

(i) Maxwell's equations for the gravito-electromagnetic field (12.12 a-d);

(ii) Relativistic equation of motion for a single particle of the beam:

$$\frac{d\mathbf{r}}{dt} = \frac{\mathbf{p}}{m_0\gamma} \; ; \quad \frac{d\mathbf{p}}{dt} = m_0\mathbf{E} + \frac{m_0}{\gamma}\mathbf{R} \times (\mathbf{B} + \mathbf{B}_0). \qquad (12.15)$$

In equations (12.12) and (12.15) $\mathbf{E}$ and $\mathbf{B}$ are the gravito-electric and — respectively — the gravito-magnetic field intensities generated by the beam, $\mathbf{B}_0$ is the constant and uniform gravito-magnetic field, $\mathbf{j} = n\frac{d\mathbf{r}}{dt}$ and n are the current density and particle density of the beam, respectively, $\mathbf{R} = \frac{\mathbf{p}}{m_0}$ is the rigidity (momentum per unit mass), and $\mathbf{p}$ is the relativistic particle momentum

$$\mathbf{p} = m_0\gamma\mathbf{v}, \qquad (12.16)$$

with

$$\gamma = \left(1 - \frac{v_\perp^2}{c^2}\right)^{-1/2}. \qquad (12.17)$$

Here m_0 is the rest mass of a single particle, $v_\perp$ is the component of the velocity orthogonal to $\mathbf{B}_0$, and γ is the relativistic Lorentz factor.

In this application, we consider a field with the following components:

(i) A gravito-electromagnetic field with harmonic space dependence of the form

$$\mathbf{E}(x,t) = \left(\Re e(E_x(t)e^{ikx}),\, \Re e(E_y(t)e^{ikx}),\, 0\right); \qquad (12.18)$$

$$\mathbf{B}(x,t) = \left(0,\, 0,\, \Re e(B(t)e^{ikx})\right), \qquad (12.19)$$

where

$$\mathbf{k} = (k, 0, 0) \qquad (12.20)$$

is the corresponding wave vector.

(ii) A stationary uniform gravito-magnetic field chosen as

$$\mathbf{B}_0 = (0, 0, B_0). \qquad (12.21)$$

Using the dimensionless variables

$$T = kct; \quad \mathbf{P} = \frac{\mathbf{p}}{m_0 c}; \quad \gamma = \sqrt{1 + P_x^2 + P_y^2};$$

$$\mathcal{B}(T) = \frac{B(t)}{kc} = \Omega_b;$$

$$\mathcal{E}_{x,y}(T) = \frac{E_{x,y}(t)}{kc^2}; \quad \Omega_B = \frac{\omega_B}{kc} = \frac{\gamma \omega_c}{kc} = \gamma \Omega_c;$$

$$\Omega_P = \frac{\omega_p}{kc} = \frac{1}{kc}\sqrt{\frac{n m_0}{\varepsilon}}, \qquad (12.22)$$

equations (12.12c,d) and (12.15) then become

$$\frac{d\mathcal{B}}{dT} = -i\mathcal{E}_y, \qquad (12.23)$$

$$\frac{d\mathcal{E}_y}{dT} = -i\mathcal{B} + \frac{1}{\gamma}\Omega_P^2 P_y e^{-iX}, \qquad (12.24)$$

$$\frac{d\mathcal{E}_x}{dT} = \frac{1}{\gamma}\Omega_P^2 P_x e^{-iX}, \tag{12.25}$$

$$\frac{dX}{dT} = \frac{P_x}{\sqrt{1 + P_x^2 + P_y^2}}, \tag{12.26}$$

$$\frac{dP_x}{dT} = \frac{1}{\gamma}\left\{\Re e\left[\mathcal{B}(T)e^{iX}\right] + \Omega_B\right\} P_y + \Re e\left[\mathcal{E}_x(T)e^{iX}\right], \tag{12.27}$$

$$\frac{dP_y}{dT} = -\frac{1}{\gamma}\left\{\Re e\left[\mathcal{B}(T)e^{iX}\right] + \Omega_B\right\} P_x + \Re e\left[\mathcal{E}_y(T)e^{iX}\right], \tag{12.28}$$

where Ω_b is the dimensionless gyro-frequency of the gravito-magnetic field generated by the gravito-electromagnetic field, Ω_B is the dimensionless cyclotron frequency generated by the stationary uniform gravito-magnetic field, Ω_c is the dimensionless relativistic cyclotron frequency generated by the stationary uniform gravito-magnetic field, and Ω_P is the dimensionless fluid "eigenfrequency".

In order to obtain the values of parameters which define an efficient mechanism of acceleration, we suppose that the motion of a single particle of the beam occurs in the gravito-electromagnetic field generated by the remaining particles of the beam. This way, the initial system with a self-consistent field is transformed into a system placed in external fields. We shall then consider the motion of a particle in a constant external gravito-magnetic field, in the presence of a transversal gravito-electromagnetic field, *i.e.*

$$\mathcal{B}(T) = \mathrm{E}_y(T) = He^{-iT}; \quad (H = const.) \tag{12.29}$$

$$\mathrm{E}_y = 0; \quad \Omega_P = 0. \tag{12.30}$$

Using a method developed in [15], the following equations describing the nonlinear dynamics of the model are obtained:

$$\frac{dX}{dT} = \frac{P_x}{\sqrt{1 + P_x^2 + P_y^2}}; \tag{12.31}$$

$$\frac{dP_x}{dT} = \Omega_c\left[\beta\cos(X - T) + 1\right] P_y \; ; \quad \beta = \frac{H}{\Omega_B} = \frac{|\Omega_b|}{\Omega_B}; \tag{12.32}$$

$$\frac{dP_y}{dT} = -\Omega_c \left[\beta \cos(X - T) + 1\right] P_x + H \cos(X - T). \qquad (12.33)$$

Since analytical solutions to the system of equations (12.31)–(12.33) are difficult to obtain, we shall integrate them numerically.

Here are the elements of the complex regular motion, for $H = 0.06$, shown in Fig. 12.1.

The numerical solutions of these equations and their correspondents with dynamics of the system are obtained by applying the fifth-order Runge-Kutta algorithm with an adaptive step size control. Let us initiate the numerical calculus for the parameters values $\Omega_B = 0.5$, $\gamma = 2$, and the initial values:

$$T = 0.001, \quad X = 0.001, \quad P_x = \sqrt{1.5}, \quad P_y = \sqrt{1.5}. \qquad (12.34)$$

In the following discussion we shall perform a complete and detailed nonlinear dynamical analysis of the system evolution described by (12.31)–(12.33): complete temporal series, Poincaré sections, complete phase space, Lyapunov exponents, bifurcation diagrams, and fractal analysis.

For small amplitudes of the gravito-electromagnetic field (*e.g.* $H = 0.06$), the particle motion is complex but retains a regular character (Figs. 12.1a–g). Then, one can associate spiral-type "trajectories" to a periodical movement having large amplitude (Fig. 12.1a) and harmonic time-increasing momentum (Figs. 12.1b,c), either in two-dimensional phase space (Figs. 12.1d–f), or in three-dimensional phase space (Fig. 12.1g).

The onset of fractalization (by means of stochastization) is observed once H exceeds 0.7 (see Figs. 12.2a–g). The debut of chaos through period doubling (Figs. 12.2b,c), as a natural result of the field — particle nonlinear interaction, initiates the particle rotation phase stochastization process in the considered gravito-magnetic field (for details see [15]). As a consequence, one can associate modified spiral-type "trajectories" to a periodic movement having small amplitude (Fig. 12.2a) and harmonic time increasing momentum, modified through period doubling (Figs. 12.2b,c), either in a two-dimensional phase space (Figs. 12,2d–f), or in a three-dimensional phase space (Fig. 12.2g).

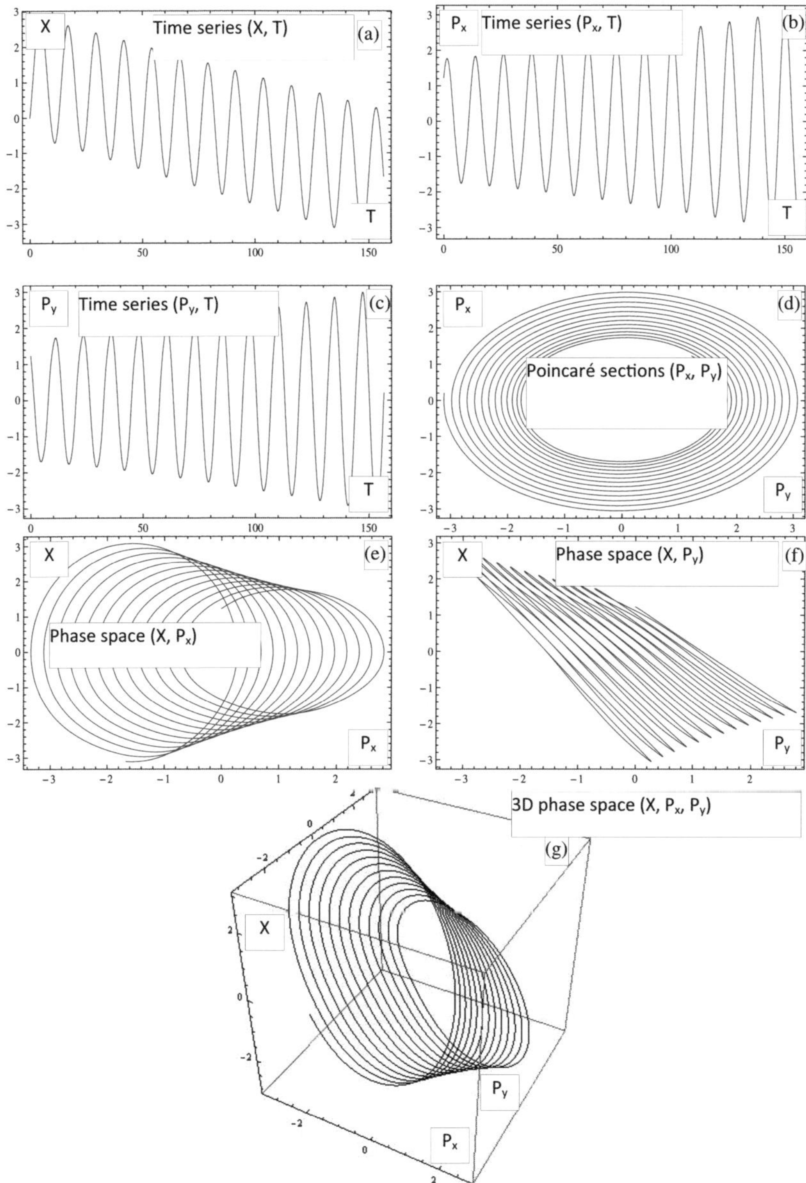

Figure 12.1. The regular characteristic of the motion. (a) time series (X, T); (b) time series (P_x, T); (c) time series (P_y, T); (d) Poincaré sections (P_x, P_y); (e) phase space (X, P_x); (f) phase space (X, P_y); (g) 3D phase space (X, P_x, P_y) for $H = 0.06$.

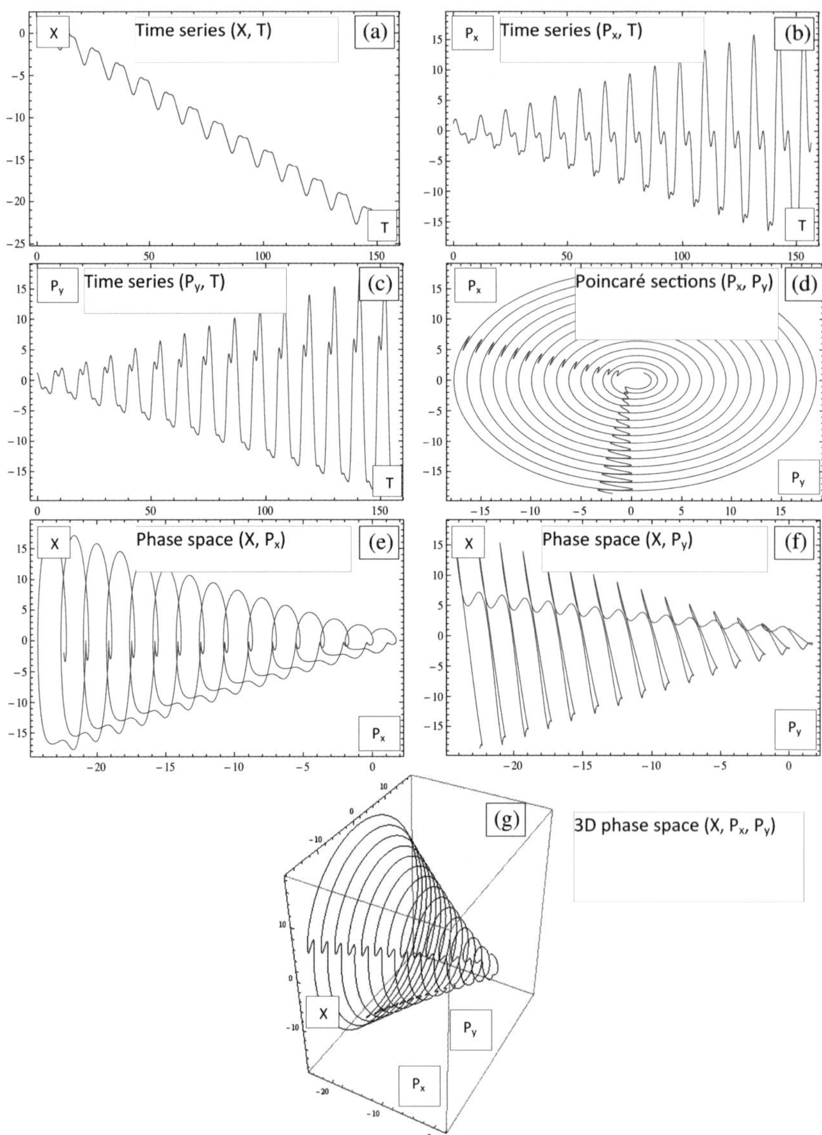

Figure 12.2.   The debut of chaos. (a) time series (X, T); (b) time series (P_x, T); (c) time series (P_y, T); (d) Poincaré sections (P_x, P_y); (e) phase space (X, P_x); (f) phase space (X, P_y); (g) 3D phase space (X, P_x, P_y) for $H = 0.7$.

Fig. 12.2, on the previous page, shows the onset of fractalization, by means of stochastization, for $H = 0.7$.

For $H = 1.4$, a gravitational gun-type effect is initiated (see Figs. 12.3a–g). The stochastization expansion process (Figs. 12.3a–c) induces the "stochastic medium" which is responsible for particle acceleration (the curve or curves connecting the "trajectories", either in a two-dimensional phase space (Figs. 12.3d–f), or in a three-dimensional phase space (Fig. 12.3g), correspond to the gravitational gun-drupe effect). The gun-effect is generated through the resonant field-particle interaction.

An extensive chaotic regime is obtained for $H = 2.7$ (see Figs. 12.4a–g). Spontaneously and unpredictably, regular behaviors intercalate with chaotic ones (emphasized either through time series (Figs. 12.4a–c), or through trajectories in a two-dimensional phase space (Figs. 12.4d–f), or, still, through "trajectories" in a three-dimensional phase space (Fig. 12.4g)). The system comes to an extended chaotic regime.

A gravitational chaotic gun-type effect erupts for $H = 3.49$ (see Figs. 12.5a–g). In the gravitational chaotic gun-type dynamics we distinguish the following sequences: at first, a localized chaotic regime emerges, subsequently a high frequency oscillation with chaotic modulation of the amplitude appears, and finally we observe the sharp, rectilinear part of the trajectory which in fact displays the gravitational gun-type effect. These sequences are emphasized either through time series (Figs. 12.5a–c), or through "trajectories" in a two-dimensional phase space (Figs. 12.5d–f), or, still, through "trajectories" in a three-dimensional phase space (Fig. 12.5g).

A gravitational multi-gun-type effect appears for $H = 4.5$ (see Figs. 12.6a–g). This is clearly illustrated either by the time series (Figs. 12.6a–c), or through "trajectories" in a two-dimensional phase space (Figs. 12.6d–f), or through "trajectories" in a three-dimensional phase space (Fig. 12.6g). The gravitational multi-gun-type effect is correlated with the jumps between different Larmor-type orbits. Vertical and large amplitude oscillations correspond to different Larmor-type orbits, while the curves connecting them correspond to the gravitational chaotic gun-type effect.

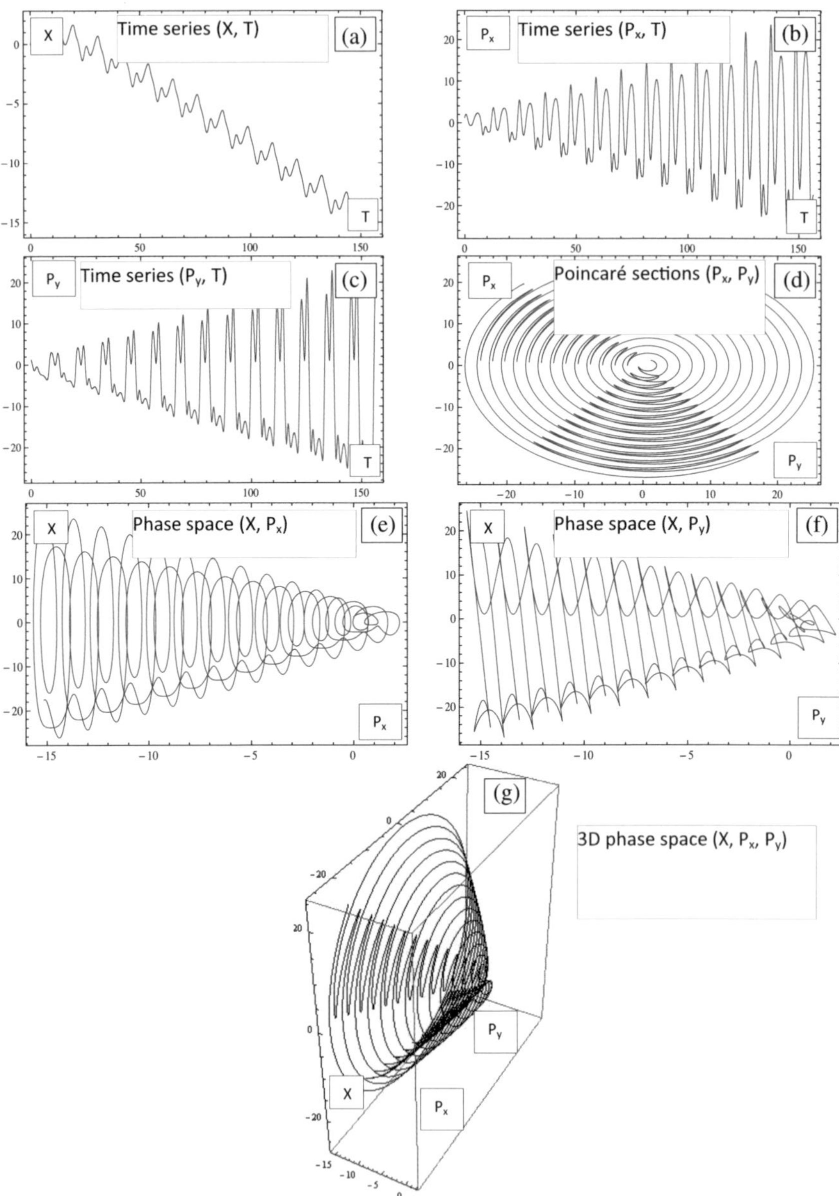

Figure 12.3. The onset of a gravitational gun-type effect for $H = 1.4$: (a) time series (X, T); (b) time series (P_x, T); (c) time series (P_y, T); (d) Poincaré sections (P_x, P_y); (e) phase space (X, P_x); (f) phase space (X, P_y); (g) 3D phase space (X, P_x, P_y).

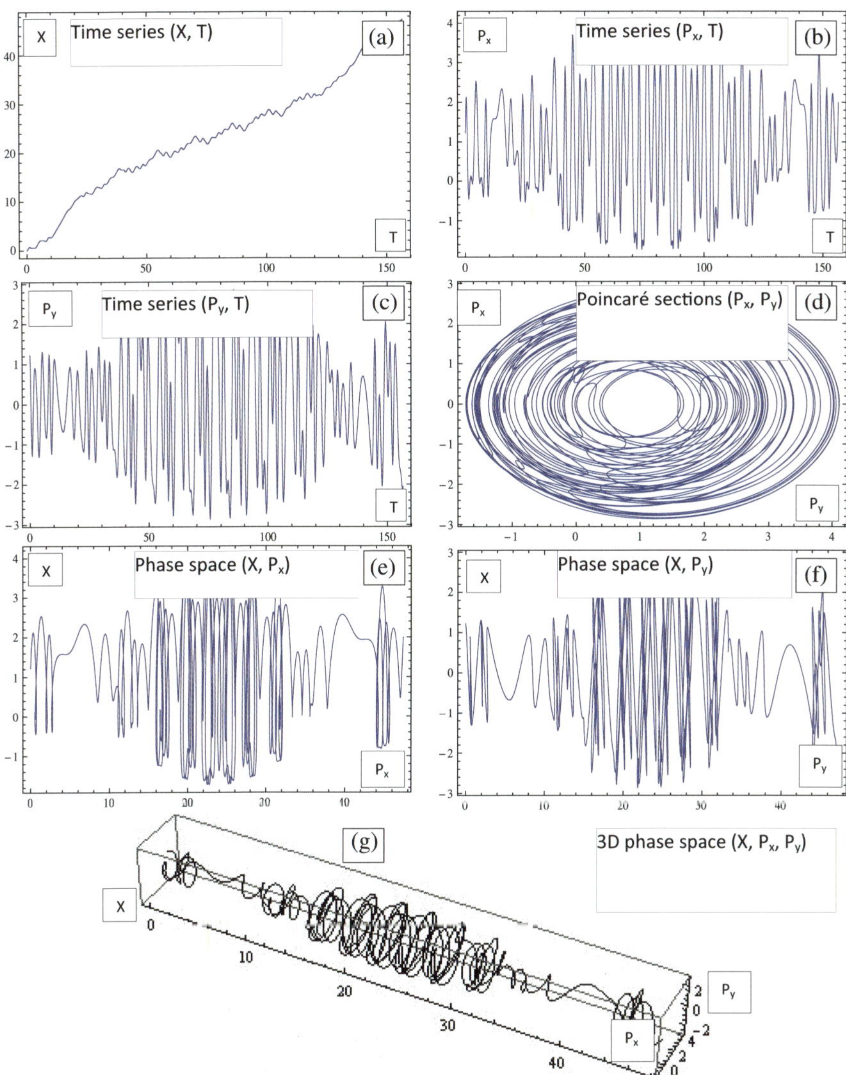

Figure 12.4. The intermittent states for $H = 2.7$: (a) time series (X, T); (b) time series (P_x, T); (c) time series (P_y, T); (d) Poincaré sections (P_x, P_y); (e) phase space (X, P_x); (f) phase space (X, P_y); (g) 3D phase space (X, P_x, P_y).

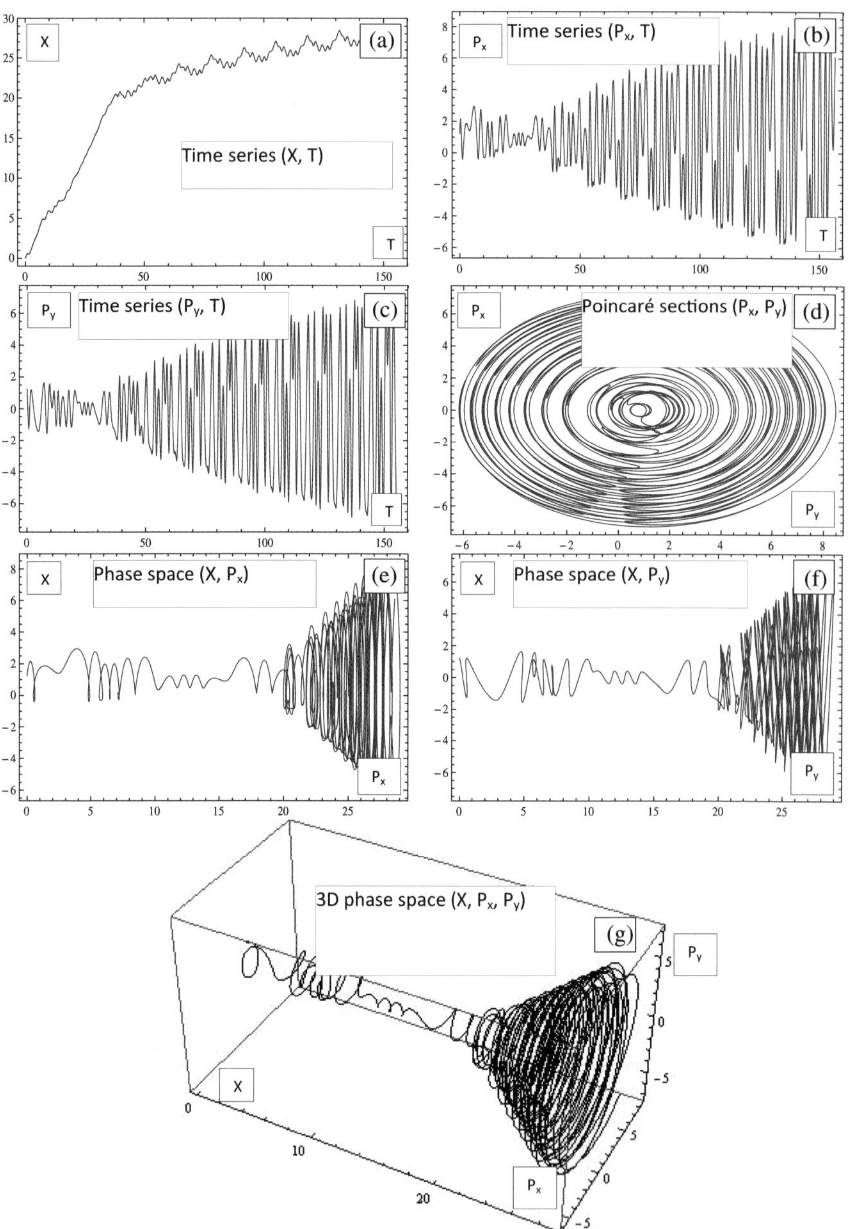

Figure 12.5. Gravitational chaotic gun-type effect for $H = 3.49$: (a) time series (X, T); (b) time series (P_x, T); (c) time series (P_y, T); (d) Poincaré sections (P_x, P_y); (e) phase space (X, P_x); (f) phase space (X, P_y); (g) 3D phase space (X, P_x, P_y).

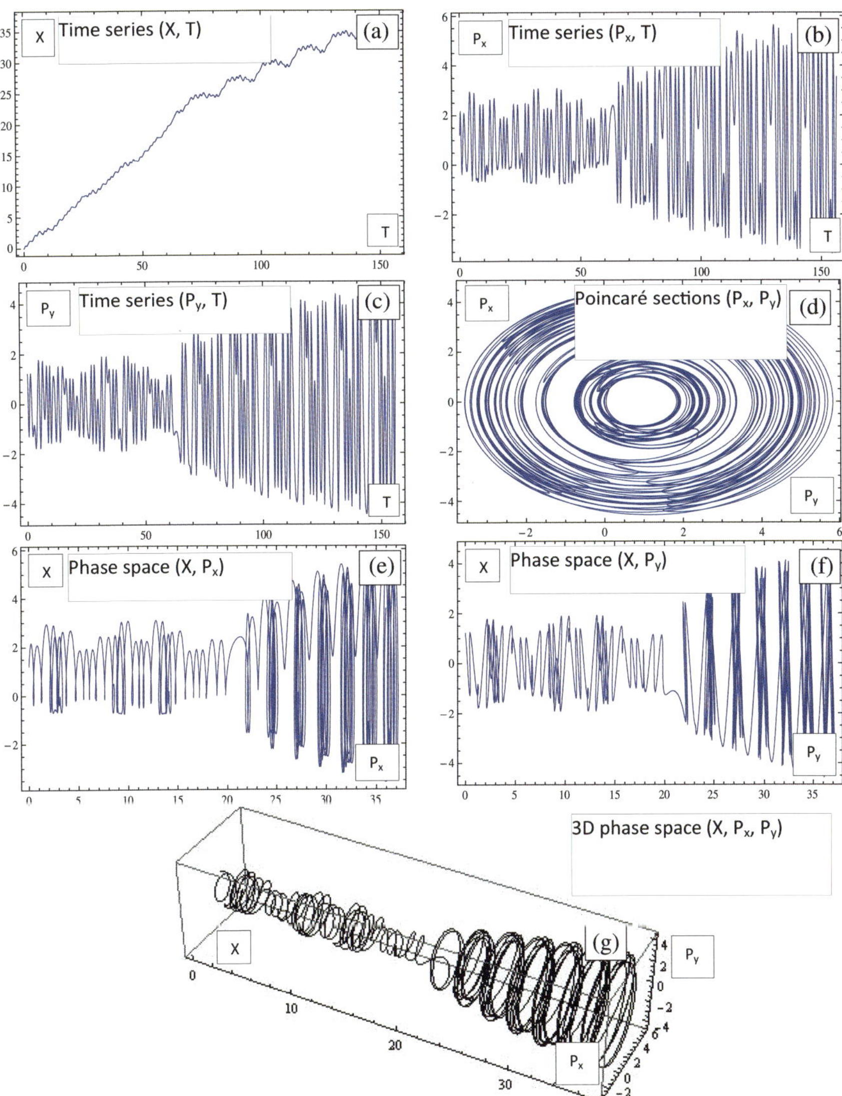

Figure 12.6. Gravitational multi-gun-type effect for $H = 4.5$: (a) time series (X, T); (b) time series (P_x, T); (c) time series (P_y, T); (d) Poincaré sections (P_x, P_y); (e) phase space (X, P_x); (f) phase space (X, P_y); (g) 3D phase space (X, P_x, P_y).

Finally, we can comment upon an interesting relation between the abstract formulas used and the associated physical reality. In this sense, a choice of type (12.29) for gravito-electromagnetic field equation is compatible with certain cosmic structures, for example with neutron stars. The gravito-magnetic field component of these material structures is:

$$B^G = 2\pi \frac{GM_N}{R_N T_N c^2} \simeq 4.6 \times 10^2 s^{-1},$$

where $G = 6.6 \cdot 10^{-11} m^3 kg^{-1} s^{-2}$ is Newton's constant, and $c = 3 \cdot 10^8 ms^{-1}$ is the speed of light in vacuum. Let us choose the following values for the rest of parameters: the star mass $M_N \simeq 10^{30} kg$, the star radius $R_N \simeq 10^4 m$, and the rotation period of the star $T_N \simeq 10^{-3} s$. With these choices, the gravito-electric field component becomes

$$E^G = \frac{GM_N}{R_N^2} \simeq 6.6 \cdot 10^{11} ms^{-2}.$$

Under these circumstances, the relativistic particles of the beam are subjected to the force field

$$F^G = cB^G \simeq 1.4 \cdot 10^{11} ms^{-2},$$

which is comparable with E^G, at least for the order of magnitude. Therefore, this type of gravito-electromagnetic field can be functional at the Observable Universe scale (for example, the white dwarf stars).

12.4. Analysis of the Lyapunov exponent

As it is known, the Lyapunov exponent dynamical system is a quantity that characterizes the rate of separation of infinitesimally close trajectories. Once the trajectories in the phase space and the laws of motion are analyzed, it comes the idea to increase the value $H = 2,5$. By decreasing the observation interval starting from $H = 4.75$, we search for zones where dynamics of our system enters a chaotic regime. It is difficult to directly observe an evolution towards chaos, because the form of circular trajectories does not change significantly (the particle returns only in some special cases). Nevertheless, in the zone $H - 4.75$ one observes a successive de-doubling of the frequencies as precursors of the routes to chaos. This analysis has been performed by studying the Fourier transform of the signal for different values of the parameters.

In order to have a more complete image of the chaotic zones, let us analyze the Lyapunov exponent in three cases, according to the modifications of initial conditions, for the three parameters X, P_x, P_y (see Figs. 12.7a–c). These graphs illustrate two-dimensional mapping of the Lyapunov exponent as a function if the two main parameters H (along the x-axis) and Ω_B (along the y-axis). One can observe a strong similarity between these three graphs, as expected.

Figures 12.7a,b,c, represent the contour diagrams for the Lyapunov exponent in terms of the parameters X, P_x and P_y, for which the initial conditions have been slightly modified. So, the values $X(0) = 0.1$, $P_x(0) = 2$, and $P_y(0) = 1$ have been considered at the initial moment. In each figure, the system evolution has been studied for small variations of about 0.1% for the initial conditions.

Figure 12.7d illustrates an analysis of the chaotic zones. The darker areas represent the zones with higher Lyapunov exponent, i.e. the chaotic zones. We can thus confirm the fact that in order to obtain extensive chaotic regimes, we must have $H > 2.5$. Thus, three important regions are outlined, and isles 2 and 3 are between $4 < H < 5$, i.e. in the same zone resulting from the previous analysis.

In the above graphs two variants of the color code are represented. In the first graph, the values for each Lyapunov coefficient level are shown, while in the second graph the color code is being displayed (see Fig. 12.8). Between the brackets one can find the open intervals for λ values, corresponding to each color.

By means of a three-dimensional representation (see Fig. 12.9), one can observe a finer scale variation of the Lyapunov exponent. The stable zones ($\lambda < 0$) were eliminated from the graph, so that the only positive values of the exponent were represented.

Let us note that the Lyapunov exponent are a rate of chaos. More precisely, it is a quantitative measure of the sensitive dependence

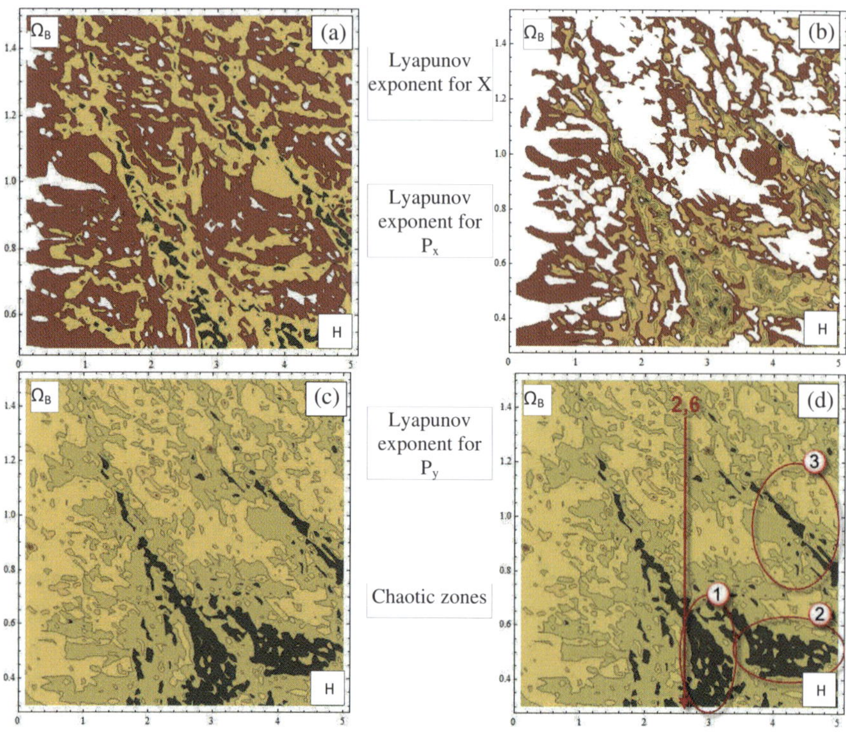

Figure 12.7.   (a)–(c): The Lyapunov exponent in three cases, according to modification of the initial conditions for the parameters X, P_x and P_y, respectively; (d) Analysis of the chaotic zones. The darker areas represent zones with higher Lyapunov exponent.

on the initial conditions, showing the rate at which the trajectory converge or diverge in the phase space. The number of Lyapunov exponents is equal to the phase space dimension. Usually, the one with the highest value is the most important because, if it is positive, then the system is characterized by a high sensitivity to initial conditions in the vicinity of the initially considered point. Also, the original Chen-Lai algorithm gradually makes chaotic an arbitrarily given discrete-time dynamical system in terms of possessing positive Lyapunov exponents with uniformly bounded orbits [16]. There are several ways of defining the Lyapunov exponents [17].

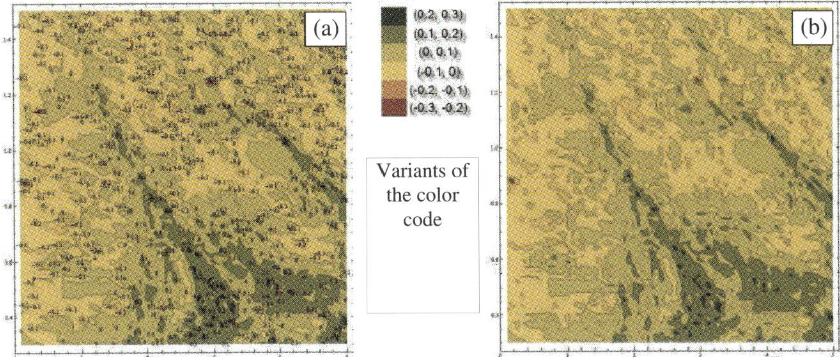

Figure 12.8.   (a)–(b). Two variants of the color code (the scale of the Lyapunov coefficients).

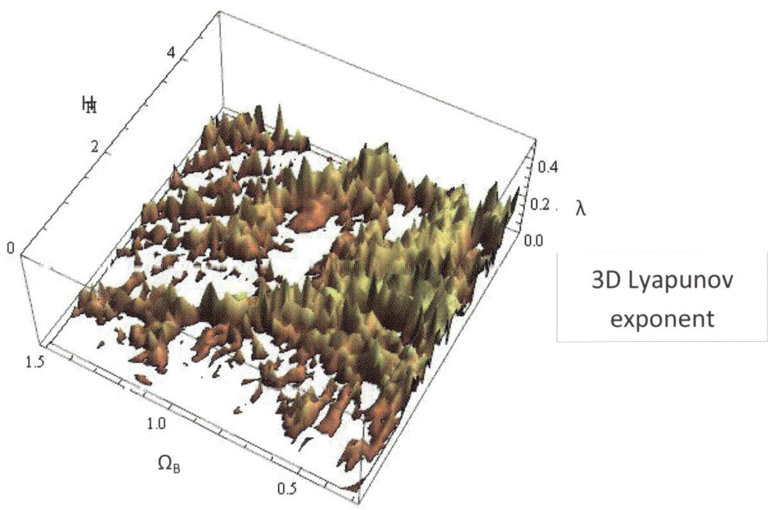

Figure 12.9.   A 3*D* representation of the Lyapunov exponent, as a function of H and Ω_B.

Here, we have used the following definition of the Lyapunov exponent:

$$\lambda = \lim_{t \to \infty} \frac{1}{t} \ln \left| \frac{\delta x(t)}{\delta x_0} \right|, \tag{12.35}$$

where $\delta x_0 = x_2(0) - x_1(0)$ is a small variation of the initial position, and $\delta x(t) = x_2(t) - x_1(t)$ is the distance between positions of the system at the moment t. Within the frame of numerical analysis in relative units we considered $t/T \approx 10^3$, where T is the specific period of the motion for $4 < H < 5$, $\Omega_B = 0.5$, and $\delta x_2(0)/x_2(0) \approx 10^{-3}$. Under these conditions, t is large enough and δx_0 is sufficiently small. Analysis for increasing precision has been performed, but not significant modification of the results were found.

12.5. Bifurcation diagram

The scenario of evolution towards chaos through resonances overlapping is accomplished: stochastic layers (inside which the particle executes random walks) surrounding the cyclotron resonances are separated from one another by invariant curves. Some of these invariant curves may vanish (intersecting resonances) and a stochastic (fractal) web-like network arises in the phase portrait, where the layers do overlap each other. The stochastic web may disappear quickly because of strong nonlinearity of the physical system. The instability is illustrated by the fact that an elliptical point converts into a hyperbolic one. Two new elliptical points of the doubled period are then formed and this represents an island-doubling bifurcation. If the energy of the particle grows continuously a cascade of successive island-doubling bifurcations (a characteristic of the Hamiltonian systems) occurs in the vicinity of the elliptical points, which form necklaces of new (smaller) stability islands corresponding to higher order resonances in the interaction of the particle with the field. When the amplitude of the field increases, the stochastic net covers the whole phase plane and even the particles with small energies can diffuse into the region of very high energies [18]. The bifurcation diagram confirms this scenario of transition to chaos (see Fig. 12.10a–c).

We observe that, in order to obtain numerical values for bifurcation diagrams, we must find the frequencies for each evolution of X, P_x, or P_y as functions of H. To this end, we can use several

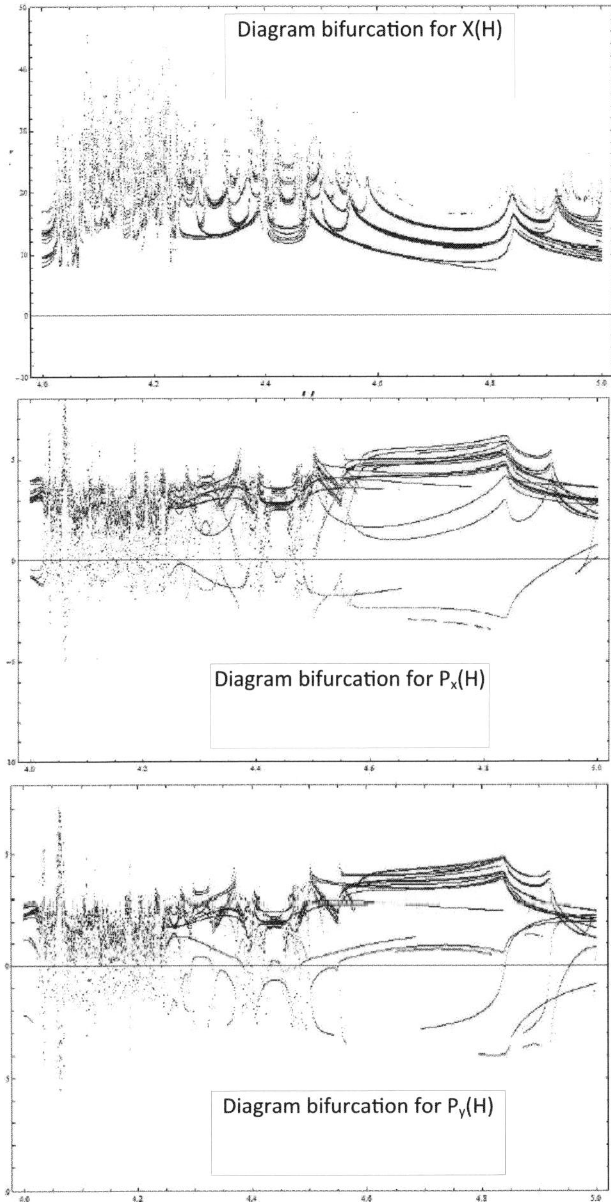

Figure 12.10. The diagrams of bifurcation by means of resonances overlapping (for $X(H)$-(a), $P_x(H)$-(b), $P_y(H)$-(c)). The sequential chaotic domain corresponds to the vertical chaotic distributions of the points.

methods. In our case, we did the mapping of X, P_x, or P_y versus time in a certain time interval. This time interval is swept along the time axis and the maxima are searched in this interval. If we find only one frequency of the same amplitude of oscillation, then, we found only one point. If we find two different frequencies, we found two different maxima and hence, the doubling of the frequencies appears. In essence, the method [19, 20] is based on the intermediate value theorem.

Only for inventory purposes we can mention here, among other efficient methods, the stochastic sensitivity function technique in the analysis of stochastic cycles and a new approach by constructing the dispersion ellipses of random trajectories for any Poincaré sections [21].

Moreover, the Fourier transform specifies the same scenario, *i.e.* the overlapping of these resonances takes place in the chaotic regime vicinity (see Fig. 12.11).

We mention that, previous to chaotic behavior, the regime is linear and the Fourier transform can be applied. But, through the same transformation, the overlapping resonances scenario is inserted.

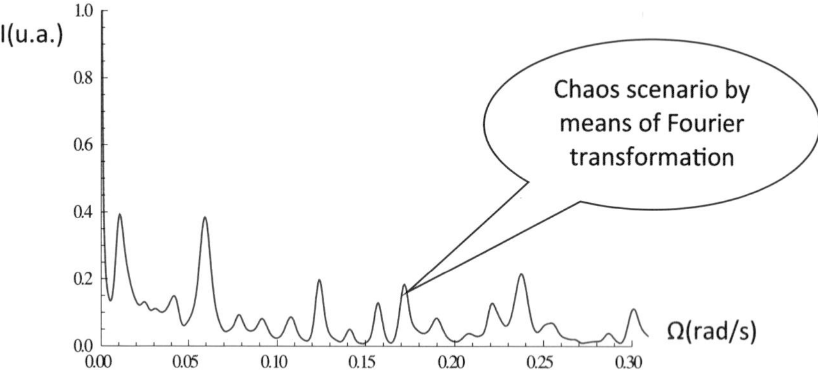

Figure 12.11.   Fourier transform specifies that the overlapping of the resonances takes place in the chaotic regime proximity.

12.6. Fractal analysis

The dynamic analysis of paragraph 3 shows not only the chaotic behavior of the system, but also the fact that particle motion is found to be a fractal one. In a previous paper [22] we also showed that the motion of particles takes place on continuous but non-differentiable curves (fractal curves) with an arbitrary and constant fractal dimension. One fundamental property of the fractal curve is self-similarity.

The fractal analysis specifies the self-similarity of the trajectories at different resolution scales, as shown in Fig. 12.12.

It then follows that the trajectory can be approximated by a fractal. A function which approximates this trajectory is:

$$X(T) = \sum_{i=1}^{n} \frac{-10 \cos(3^i T)}{\sqrt{i}} + 10\,nT. \tag{12.36}$$

In Fig. 12.13 this function is displayed at different resolution scales, either in the presence of a guide (a)–(d), or in its absence (e).

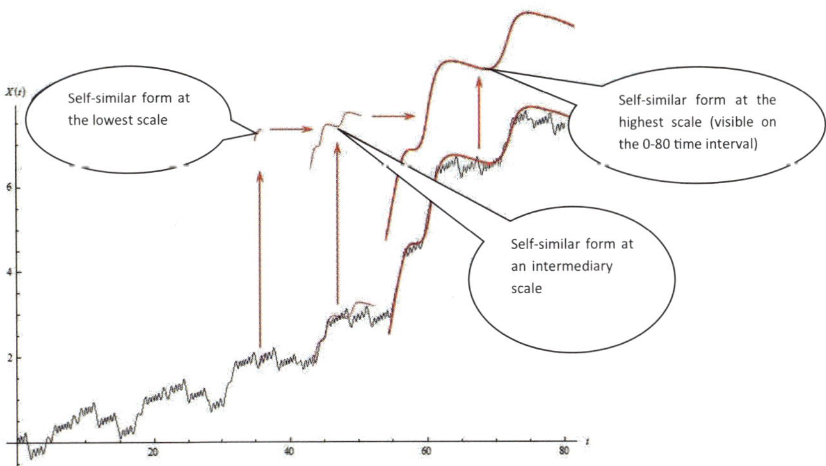

Figure 12.12.   The self-similarity of the trajectories at different resolution scales for $\Omega_B = 1.52, H = 4.471, \beta = 8.942$.

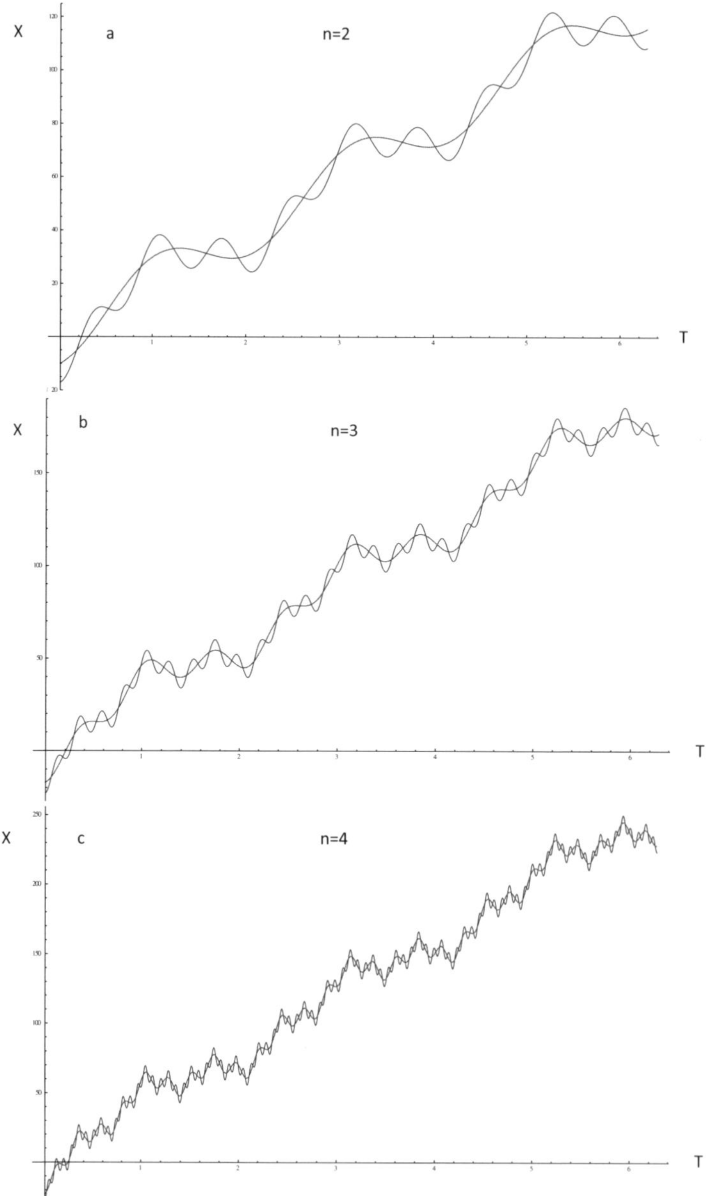

Figure 12.13. Function (12.36) at different resolution scales in the presence of a guide (a)–(d), and in its absence (e).

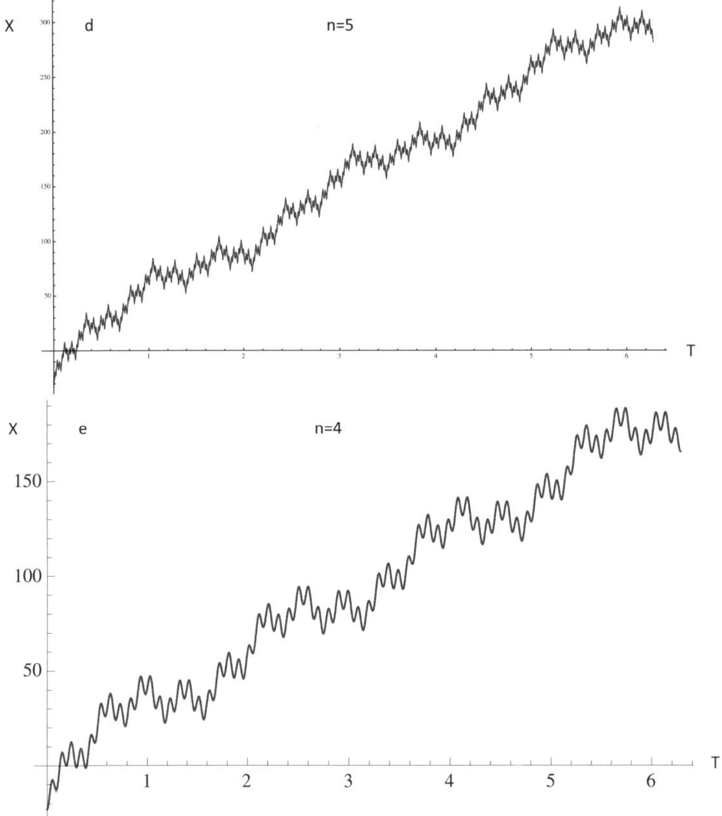

Figure 12.13. (*Continued*)

Taking into account that (12.36) is an alternative to the Weierstrass function, the fractal curve dimension is $D_f = 1.5$. A new study of nearly one million galaxies suggests matter in the universe is distributed in a fractal pattern (at any scale), with the fractal dimension going asymptotically to 1.5 [23].

We can conclude, therefore, that in the above investigations a good coherence between the theoretical and experimental model is achieved. As a consequence, the trajectories of motion of the neutral particle as a result of its nonlinear interaction with a field are described by a fractal curve [24].

12.7. Conclusions

The nonlinear dynamics analysis (temporal series, Poincaré sections, phase space, Lyapunov exponents, bifurcation diagrams, etc.) of the particle-field nonlinear interactions (composite fields containing a gravito-electromagnetic field and a static magnetic field) is extensively presented. Some physical mechanisms (gravitational gun-type, chaotic gun-type and multi-gun-type effects) can well explain the transition of a test particle from one orbit to another (classical analogue of quantum absorption). The fractal analysis of the trajectories resulting from the particle-field nonlinear interaction shows that these curves possess the property of continuity and non-differentiability, i.e. they are fractal curves. In such a context, we were able to apply the scale relativity theory to the study of the stationary states of the test particle, introducing the quantization at cosmological scale. Therefore, the trajectory fractalization represents a natural mechanism of quantization at the cosmological scale.

References (Chapter 12)

[1] Strogatz, S. H. [2001] *Nonlinear Dynamics and Chaos: With Applications to Physics, Biology, Chemistry, and Engineering*, (Westview Press, Boulder, Colorado).

[2] Mandelbrot, B. B. [2004] *Fractals and Chaos: The Mandelbrot Set and Beyond*, (Springer, New York).

[3] Ott, E. [2002] *Chaos in Dynamical Systems*, 2nd Ed. (Cambridge University Press, New York).

[4] Muthuswamy, B. and Chua, L. O. [2010] *Simplest chaotic circuit*, Int. J. Bifurcation and Chaos 20, 1567–1580.

[5] Nayfeh, A. H. [2000] *Nonlinear Interactions: Analytical, Computational, and Experimental Methods*, (Wiley-Interscience, New York).

[6] Laskar, J. [1989] *A numerical experiment on the chaotic behavior of the solar-system*, Nature 338, 237–238.

[7] Sussman, G. J. and Wisdom, J. [1992] *Chaotic evolution of the solar-system*, Science 257, 56–62.

[8] Laskar, J. [1994] it Large-scale chaos in the Solar System, Astron. Astrophys. 287, L9–L12.

[9] Argyle, B. [2004] *Observing and Measuring Double Stars* (Springer, London).

[10] Iping, R. C., Sonneborn, G., Gull, T. R., Massa, D. L. and Hillier, D. J. [2005] *Detection of a Hot Binary Companion of eta Carinae*, Astrophys. J. 633, L37–L40.

[11] Peng, H. and Qin, C. B. [1984] *Precession of a gyroscope on the ground and the existence of magnetic-type gravitation*, Phys. Lett. A 103, 197–199.

[12] Peng, H. [1990] *A new approach to studying local gravitomagnetic effects on a superconductor*, Gen. Rel. and Gravit. 22, 609–617.

[13] Ciubotariu, C. [1991] *Absorption of gravitational-waves*, Phys. Lett. A 158, 27–30.

[14] Agop, M., Buzea, C. G. and Nica, P. [2000] *Local gravito-electromagnetic effects on a superconductor*, Physica C 339, 120–128.

[15] Argyris, J. and Ciubotariu, C. [2000] *A chaotic gun effect for relativistic charged particles*, Chaos Solit. Fract. 11, 1001–1014.

[16] Lai, D. and Chen, G. [2006] *Distribution of controlled Lyapunov exponents via the Lai-Chen algorithm*, Comput. Math. Appl. 52, 1649–1656.

[17] Elhadj, Z. and Sprott, J. C. [2009] *Some explicit formulas of Lyapunov exponents for three-dimensional quadratic mappings*, Front. Phys. China 4, 549–555.

[18] Antici, A., Maric, C. and Agop, M. [2009] *Chaos Through Stochastization*, (Ars Longa, Iasi).

[19] Mandelbrot, B. B. [1983] *The Fractal Geometry of Nature*, (W.H. Freeman and Company, New York).

[20] Gouyet, J. F. [1996] *Physics and Fractal Structures*, (Springer, Berlin).

[21] Bashkirtseva, I., Chen, G. and Ryashko, L. [2010] *Analysis of stochastic cycles in the Chen system*, Int. J. Bifurcation and Chaos 20, 1439–1450.

[22] Munceleanu, G. V., Paun, V. P., Casian-Botez, I. and Agop, M. [2011] *The microscopic-macroscopic scale transformation through a chaos scenario in the fractal space-time theory*, Int. J. Bifurcation and Chaos 21, 603–618.

[23] Nottale, L. [2011] *Scale Relativity and Fractal Space-Time: A New approach to unifying relativity and quantum mechanics*, (Imperial College Press, London).

[24] Nottale, L. [2010] *Scale Relativity and Fractal Space-Time: Theory and Applications*, Found. Sci. 15, 101–152.

Chapter 13

Fractality at Small Scale. Fractal Model of the Atom

13.1. Introduction

The theoretical description of microphysical systems is based on the wave mechanics of Schrödinger [1], the matrix mechanics of Heisenberg [2], or the path-integral mechanics of Feynman [3]. Another approach is the hydrodynamic formulation of quantum mechanics due to Madelung [4], de Broglie [5], Takabayasi [6] and Bohm [7] (idea of 'subquantum medium'). The hydrodynamic theory of quantum mechanics has been later extended by de Broglie (the idea of 'double solution') ([8], [9]).

The theory of scale relativity (SR) is build by completing the standard laws of classical physics (motion in space-time) by new scale laws (in which "the space-time resolutions are used as intrinsic variables, playing for scale transformations the same role as played by velocities for motion transformations" ([10]–[12]). This model is based both on the fractal space-time concept, that is also introduced by Ord [13] and Weibel et al. [14] and on a generalization of Einstein's principle of relativity to scale transformations. The space-time resolutions are redefined as characterizing the state of scale of reference systems, in the same way as velocity characterizes their state of motion. Then, one requires that the laws of physics apply for any state of the reference system, of motion (principle of motion-relativity) and of scale (principle of scale-relativity). Mathematically, the principle of SR is achieved by the principle of scale-covariance, requiring that the equations of physics keep their simplest form under transformations of resolution ([10]–[12]).

Three scales of interaction of SRT were developed: (i) A 'Galilean' version corresponding to the standard fractals with constant fractal dimensions [15] and which involves quantum mechanics ([10]–[12]); (ii) A special scale-relativistic version which implies the high energy physics ([10]–[12], [16]); (iii) A 'general scale-relativistic' version which implies the cosmology ([10]–[12], [17]).

13.2. Nottale's model of the scale relativity theory

In Nottale's model of SRT it is supposed that the motion of "microparticles" takes place on fractal (continuous but non-differentiable) curves of fractal dimension D_F. A manifold compatible with such motion was called fractal space-time. The fractal nature of space-time implied, through non-differentiability, the following elements ([10]–[12]): (i) since the number of geodesics is infinite, a statistical (fluid-like) description was accepted; (ii) each geodesic was considered a fractal curve. As a consequence, new terms were added in the differential equations of motion; (iii) the breaking of differential time reflection invariance ([10], [16]). So, the usual definitions of the derivative of a given function with respect to time ([10], [12])

$$\frac{df}{dt} = \lim_{\Delta t \to 0+} \frac{f(t + \Delta t) - f(t)}{\Delta t} = \lim_{\Delta t \to 0-} \frac{f(t) - f(t - \Delta t)}{\Delta t}$$

are equivalent in the differentiable case. One passes from one to the other by the transformation $\Delta t \to -\Delta t$ (time reflection invariance at the infinitesimal level). In the non-differentiable case, two functions (df_+/dt) and (df_-/dt) are defined as explicit functions of t and of dt ([10], [12]),

$$\frac{df_+}{dt} = \lim_{\Delta t \to 0+} \frac{f(t + \Delta t, \Delta t) - f(t, \Delta t)}{\Delta t}$$

$$\frac{df_-}{dt} = \lim_{\Delta t \to 0-} \frac{f(t, \Delta t) - f(t - \Delta t, \Delta t)}{\Delta t}.$$

$$(13.1a,b)$$

The sign $(+)$ corresponds to the forward process and $(-)$ to the backward process.

By applying (13.1a,b) to the forward and backward position vectors, $d\mathbf{X}_\pm$, we have (for details see [10,12])

$$d\mathbf{X}_\pm = d\mathbf{x}_\pm + d\boldsymbol{\xi}_\pm = \mathbf{v}_\pm dt + d\boldsymbol{\xi}_\pm, \qquad (13.1\mathrm{c,d})$$

with $d\mathbf{x}_\pm$ the forward and backward mean position vectors, $\mathbf{v}_\pm$ the forward and backward mean velocity vectors

$$\mathbf{v}_+ = \frac{d\mathbf{x}_+}{dt} = \lim_{\Delta t \to 0+} \left\langle \frac{\mathbf{X}(t + \Delta t) - \mathbf{X}(t)}{\Delta t} \right\rangle;$$

$$\mathbf{v}_- = \frac{d\mathbf{x}_-}{dt} = \lim_{\Delta t \to 0-} \left\langle \frac{\mathbf{X}(t) - \mathbf{X}(t - \Delta t)}{\Delta t} \right\rangle, \qquad (13.2\mathrm{a,b})$$

and $d\boldsymbol{\xi}$ a measure of non-differentiability (a fluctuation induced by the fractal properties of trajectory) with the average value

$$\langle d\boldsymbol{\xi}_\pm \rangle = 0. \qquad (13.3)$$

While the classic concept of velocity is unique, if the space-time has a fractal structure two velocities ($\mathbf{v}_+$ and $\mathbf{v}_-$) are introduced instead. This "two valueness" of the velocity vector is a new, specific consequence of non-differentiability that has no standard counterpart (in the sense of differential physics) ([10], [12]).

Since $\mathbf{v}_+$ and $\mathbf{v}_-$ have the same relevance, the only solution is to consider both the forward ($dt > 0$) and backward ($dt < 0$) processes together. Then, the complex velocity is defined as ([10], [12]):

$$\mathbf{V} = \frac{\mathbf{v}_+ + \mathbf{v}_-}{2} - i\frac{\mathbf{v}_+ - \mathbf{v}_-}{2} = \frac{d\mathbf{x}_+ + d\mathbf{x}_-}{2dt} - i\frac{d\mathbf{x}_+ - d\mathbf{x}_-}{2dt}. \qquad (13.4)$$

If $(\mathbf{v}_+ + \mathbf{v}_-)/2$ can be considered as differentiable (classical) velocity, then the difference $(\mathbf{v}_+ - \mathbf{v}_-)/2$ is the non-differentiable velocity.

Using the notation $d\mathbf{x}_{\pm} = d_{\pm}\mathbf{x}$, relation (13.4) becomes

$$\mathbf{V} = \left(\frac{d_+ + d_-}{2dt} - i\frac{d_+ - d_-}{2dt} \right) \mathbf{x}. \tag{13.5}$$

Following this reasoning, it was Nottale's idea to define the operator

$$\frac{\delta}{dt} = \frac{d_+ + d_-}{2dt} - i\frac{d_+ - d_-}{2dt}. \tag{13.6}$$

We note that Nottale's papers ([16], [17]) refer mainly to the physical background of the scale relativity theory in connection with quantum mechanics, i.e. second order terms in the equation of motion and the fractal curves of dimension $D_F = 2$. This implies only movements on geodesics of "Schrödinger type". In the following investigations, Nottale's model shall be extended by considering arbitrary constant fractal dimension curves and third order terms in the equation of motion.

13.3. Extended model of the scale relativity

Let us now assume that the curve describing the movement (continuous, but non-differentiable) is immersed in a 3-dimensional space, and that $\mathbf{X}$ of components X^i $(i = \overline{1,3})$ is the position vector of a point on the curve. Let us also consider a function $f(\mathbf{X}, t)$ and expand its total differential up to the third order term:

$$df = \frac{\partial f}{\partial t}dt + \nabla f \cdot d\mathbf{X} + \frac{1}{2}\frac{\partial^2 f}{\partial X^i \partial X^j}d_{\pm}X^i d_{\pm}X^j$$

$$+ \frac{1}{6}\frac{\partial^3 f}{\partial X^i \partial X^j \partial X^k}d_{\pm}X^i d_{\pm}X^j d_{\pm}X^k. \tag{13.7}$$

Only the first three terms of (13.7) were used in Nottale's theory (i.e. second order terms in the equation of motion) ([10], [12]).

Denoting $dX^i_\pm = d_\pm X^i$, the forward and backward average values of this relation writes

$$\langle d_\pm f \rangle = \left\langle \frac{\partial f}{\partial t} dt \right\rangle + \langle \nabla f \cdot d_\pm \mathbf{X} \rangle + \frac{1}{2} \left\langle \frac{\partial^2 f}{\partial X^i \partial X^j} d_\pm X^i d_\pm X^j \right\rangle$$

$$+ \frac{1}{6} \left\langle \frac{\partial^3 f}{\partial X^i \partial X^j \partial X^k} d_\pm X^i d_\pm X^j d_\pm X^k \right\rangle . \tag{13.8a,b}$$

From now on, we shall assume that the mean values of the function f and its derivatives coincide with themselves, and the differentials $d_\pm X^i$ and dt are independent, therefore the average of their product coincides with the product of averages. Thus equations (13.8a,b) become:

$$d_\pm f = \frac{\partial f}{\partial t} dt + \nabla f \cdot \langle d_\pm \mathbf{X} \rangle + \frac{1}{2} \frac{\partial^2 f}{\partial X^i \partial X^j} \left\langle d_\pm X^i d_\pm X^j \right\rangle$$

$$+ \frac{1}{6} \frac{\partial^3 f}{\partial X^i \partial X^j \partial X^k} \left\langle d_\pm X^i d_\pm X^j d_\pm X^k \right\rangle . \tag{13.9a,b}$$

or, using equations (13.1c,d) and the property (13.3),

$$d_\pm f = \frac{\partial f}{\partial t} dt + \nabla f \cdot d_\pm \mathbf{X} + \frac{1}{2} \frac{\partial^2 f}{\partial X^i \partial X^j} \left(d_\pm x^i d_\pm x^j + \left\langle d\xi^i_\pm d\xi^j_\pm \right\rangle \right)$$

$$+ \frac{1}{6} \frac{\partial^3 f}{\partial X^i \partial X^j \partial X^k} \left(d_\pm x^i d_\pm x^j d_\pm x^k + \left\langle d\xi^i_\pm d\xi^j_\pm d\xi^k_\pm \right\rangle \right) . \tag{13.10a,b}$$

Since $d\xi^i_\pm$ describes the fractal properties of the fractal curve which has the fractal dimension D_F (the mathematical formalism of scale relativity is independent on the way how the fractal dimension is defined — for details see [15]) — but once this definition is chosen it should be maintained), through a convenient choice both of the "time scale", τ, and of the "length scale", λ, it is natural to impose that the dimensionless fluctuation to D_F power, $(d\xi^i_\pm/\lambda)^{D_F}$ equals the dimensionless time parameter, (dt/τ), *i.e.*

$$d\xi_\pm = \lambda \left(\frac{dt}{\tau}\right)^{1/D_F}. \tag{13.11}$$

Significance of the time-quantities t and τ results from the Random Walk (Brownian motion) or its Levy generalization motion ([10]–[12]). In the SRT the differential time dt is identified with the time resolution $dt = \delta t$ (substitution principle ([10]–[12]) while τ corresponds to the fractal — non-fractal transition time.

Even if the average value of the fractal coordinate $d\xi_\pm^i$ is null (see (13.3)), for higher order of the fractal coordinate average the situation can be different. First, let us focus on the mean $\langle d\xi_\pm^i d\xi_\pm^j \rangle$. If $i \neq j$, this averaged quantity is zero, due to independence of $d\xi^i$ and $d\xi^j$. So, using (13.11). we can write

$$\left\langle d\xi_\pm^i d\xi_\pm^j \right\rangle = \delta^{ij} \frac{\lambda^2}{\tau} \left(\frac{dt}{\tau}\right)^{\left(\frac{2}{D_F}\right)-1} dt, \tag{13.12a, b}$$

with

$$\delta^{ij} = \begin{cases} 1, & \text{if } i = j \\ 0, & \text{if } i \neq j, \end{cases}$$

and we have assumed that

$$\begin{cases} \langle d\xi_+^i d\xi_+^j \rangle > 0 & \text{and} \quad dt > 0; \\ \langle d\xi_-^i d\xi_-^j \rangle > 0 & \text{and} \quad dt < 0. \end{cases}$$

Next, let us consider $\langle d\xi_\pm^i d\xi_\pm^j d\xi_\pm^k \rangle$. If $i \neq j \neq k$, this average quantity is zero, due to the mutual independence of $d\xi^i$, $d\xi^j$ and $d\xi^k$. Using equation (13.11), we can write

$$\langle d\xi_\pm^i d\xi_\pm^j d\xi_\pm^k \rangle = \delta^{ijk} \frac{\lambda^3}{\tau} \left(\frac{dt}{\tau}\right)^{\left(\frac{3}{D_F}\right)-1} dt \tag{13.13a, b}$$

with

$$\delta^{ijk} = \begin{cases} 1 & \text{if} \quad i = j = k; \\ 0 & \text{if} \quad i \neq j \neq k, \end{cases}$$

and we have assumed that

$$\begin{cases} \langle d\xi_+^i\, d\xi_+^j\, d\xi_+^k \rangle > 0 \quad \text{and} \quad dt > 0; \\[2mm] \langle d\xi_-^i\, d\xi_-^j\, d\xi_-^k \rangle < 0 \quad \text{and} \quad dt < 0. \end{cases}$$

Then, equations (13.10a,b) can be written as

$$d_\pm f = \frac{\partial f}{\partial t} dt + \nabla f \cdot d_\pm \mathbf{x} + \frac{1}{2}\frac{\partial^2 f}{\partial X^i \partial X^j} d_\pm x^i d_\pm x^j$$

$$\pm \frac{\partial^2 f}{\partial X^i \partial X^j} \delta^{ij} \frac{\lambda^2}{2\tau} \left(\frac{dt}{\tau}\right)^{\left(\frac{2}{D_F}\right)-1} dt$$

$$+ \frac{1}{6}\frac{\partial^3 f}{\partial X^i \partial X^j \partial X^k} d_\pm x^i d_\pm x^j d_\pm x^k$$

$$+ \frac{\partial^3 f}{\partial X^i \partial X^j \partial X^k} \delta^{ijk} \frac{\lambda^3}{6\tau} \left(\frac{dt}{\tau}\right)^{\left(\frac{3}{D_F}\right)-1} dt. \qquad (13.14\text{a,b})$$

Dividing by dt and neglecting the terms containing differential factors (for details on the method see [10], [18]), equations (13.14a,b) are reduced to:

$$\frac{d_\pm f}{dt} = \frac{\partial f}{\partial t} + \mathbf{v}_\pm \cdot \nabla f \pm \frac{\lambda^2}{2\tau}\left(\frac{dt}{\tau}\right)^{\left(\frac{2}{D_F}\right)-1}\Delta f$$

$$+ \frac{\lambda^3}{6\tau}\left(\frac{dt}{\tau}\right)^{\left(\frac{3}{D_F}\right)-1}\nabla^3 f, \qquad (13.15\text{a,b})$$

where $\nabla^3 = \sum_i \partial^3/\partial X_i^3$. These relations allow us to define the operator

$$\frac{d_\pm}{dt} = \frac{\partial}{\partial t} + \mathbf{v}_\pm \cdot \nabla \pm \frac{\lambda^2}{2\tau}\left(\frac{dt}{\tau}\right)^{\left(\frac{2}{D_F}\right)-1}\Delta$$

$$+ \frac{\lambda^3}{6\tau}\left(\frac{dt}{\tau}\right)^{\left(\frac{3}{D_F}\right)-1}\nabla^3. \qquad (13.16\text{a,b})$$

Under these circumstances, let us calculate $(\delta f/dt)$. Taking into account equations (13.5), (13.6), and (13.16a,b), we obtain:

$$\frac{\delta f}{dt} = \frac{1}{2}\left[\frac{d_+f}{dt} + \frac{df_-}{dt} - i\left(\frac{d_+f}{dt} - \frac{df_-}{dt}\right)\right] = \frac{\partial f}{\partial t} + \mathbf{V}\cdot\nabla f$$

$$- i\frac{\lambda^2}{2\tau}\left(\frac{dt}{\tau}\right)^{\left(\frac{2}{D_F}\right)-1}\Delta f + \frac{\lambda^3}{6\tau}\left(\frac{dt}{\tau}\right)^{\left(\frac{3}{D_F}\right)-1}\nabla^3 f. \qquad (13.17)$$

This relation allows us to define the fractal operator

$$\frac{\delta}{dt} = \frac{\partial}{\partial t} + \mathbf{V}\cdot\nabla - i\frac{\lambda^2}{2\tau}\left(\frac{dt}{\tau}\right)^{\left(\frac{2}{D_F}\right)-1}\Delta$$

$$+ \frac{\lambda^3}{6\tau}\left(\frac{dt}{\tau}\right)^{\left(\frac{3}{D_F}\right)-1}\nabla^3. \qquad (13.18)$$

Let us now apply the principle of scale covariance and postulate that the transition from classical (differentiable) mechanics to the "fractal" mechanics which is considered here can be implemented by replacing the standard time derivative d/dt by the complex operator δ/dt (as a result of generalization of the principle of scale covariance given by Nottale in [10]–[12]). As a consequence, we are now able to write the equation of geodesics (a generalization of the Newton's first principle) in a fractal space-time under its covariant form:

$$\frac{\delta \mathbf{V}}{dt} = \frac{\partial \mathbf{V}}{\partial t} + \mathbf{V}\cdot\nabla\mathbf{V} - i\frac{\lambda^2}{2\tau}\left(\frac{dt}{\tau}\right)^{\left(\frac{2}{D_F}\right)-1}\Delta\mathbf{V}$$

$$+ \frac{\lambda^3}{6\tau}\left(\frac{dt}{\tau}\right)^{\left(\frac{3}{D_F}\right)-1}\nabla^3\mathbf{V} = 0. \qquad (13.19)$$

This relation shows that the global complex acceleration field $\delta\mathbf{V}/dt$ depends on the local complex acceleration field $\partial_t\mathbf{V}$, the non-linearity (convective term $\mathbf{V}\cdot\nabla\mathbf{V}$), the dissipative term $\Delta\mathbf{V}$, and on the dispersive term $\nabla^3\mathbf{V}$ as well.

If the motions of the fractal fluid are irrotational, *i.e.* $\Omega = \nabla \times \mathbf{V} = 0$, we can choose $\mathbf{V}$ of the form

$$\mathbf{V} = \nabla \phi. \tag{13.20}$$

where ϕ is a complex velocity potential. Equation (13.19) then becomes

$$\frac{\partial \mathbf{V}}{\partial t} + \nabla \left(\frac{\mathbf{V}^2}{2} \right) - i \frac{\lambda^2}{2\tau} \left(\frac{dt}{\tau} \right)^{\left(\frac{2}{D_F} \right)-1} \Delta \mathbf{V}$$

$$+ \frac{\lambda^3}{6\tau} \left(\frac{dt}{\tau} \right)^{\left(\frac{3}{D_F} \right)-1} \nabla^3 \mathbf{V} = 0. \tag{13.21}$$

Substituting (13.20) in (13.21) and integrating, we also have

$$\frac{\partial \phi}{\partial t} + \frac{1}{2}(\nabla \phi)^2 - i \frac{\lambda^2}{2\tau} \left(\frac{dt}{\tau} \right)^{\left(\frac{2}{D_F} \right)-1} \Delta \phi$$

$$+ \frac{\lambda^3}{6\tau} \left(\frac{dt}{\tau} \right)^{\left(\frac{3}{D_F} \right)-1} \nabla^3 \phi = F(t), \tag{13.22}$$

where $F(t)$ is a function of time only. We note that equation (13.21) has been reduced to a single scalar relation (13.22), which is a generalized Bernoulli (GB)-type equation.

Let us choose the complex velocity potential ϕ in the form

$$\Phi = -i \frac{\lambda^2}{2\tau} \left(\frac{dt}{\tau} \right)^{\left(\frac{2}{D_F} \right)-1} \ln \psi, \tag{13.23}$$

where ψ behaves both as a velocity potential and a wave function ([10], [16]–[18]). Then ψ, in view of (13.22), satisfies a generalized Schrödinger (GS)-type equation:

$$\frac{\lambda^4}{4\tau^2} \left(\frac{dt}{\tau} \right)^{\left(\frac{4}{D_F} \right)-2} \Delta \psi + i \frac{\lambda^2}{2\tau} \left(\frac{dt}{\tau} \right)^{\left(\frac{2}{D_F} \right)-1} \partial_t \psi$$

$$+ \left(i \frac{\lambda^5}{12\tau^2} \left(\frac{dt}{\tau} \right)^{\left(\frac{5}{D_F} \right)-2} (\nabla^3 \ln \psi) + \frac{F(t)}{2} \right) \psi = 0. \tag{13.24}$$

13.4. The dissipative approximation of motion in fractal structures. Fractal model of the atom

Let us consider the following assumptions: (i) the microparticle motions take place on fractal curves with the fractal dimension $D_F = 2$; (ii) The ratio (dt/τ) satisfies the condition $(dt/\tau) \ll 1$. Then, according to equation (13.21), the microparticle movement is described by a generalized Navier-Stokes (GNS) type equation,

$$\frac{\delta \mathbf{V}}{dt} = \frac{\partial \mathbf{V}}{\partial t} + \mathbf{V} \cdot \nabla \mathbf{V} - i\frac{\lambda^2}{2\tau}\Delta \mathbf{V} = 0, \tag{13.25}$$

with an imaginary viscosity coefficient $\eta = i(\lambda^2/2\tau)$. This means that in the fractal space-time the dissipative and convective effects are dominant in comparison with the dispersive ones. By substituting in equation (13.25) the relations (13.20) and (13.23), up to an arbitrary phase factor which may be set to zero by a suitable choice of the phase ψ, a Schrödinger-type equation is obtained

$$\frac{\lambda^4}{4\tau^2}\Delta\psi + i\frac{\lambda^2}{2\tau}\partial_t\psi = 0. \tag{13.26a}$$

For $\psi = \sqrt{\rho}e^{iS}$, with $\sqrt{\rho}$ the amplitude and S the phase of ψ, the complex velocity field (13.5), in view of (13.20) and (13.23), becomes

$$\mathbf{V} = \mathbf{v} + i\mathbf{u}; \quad \mathbf{v} = \frac{\lambda^2}{\tau}\nabla S; \quad \mathbf{u} = -\frac{\lambda^2}{2\tau}\nabla(\ln\rho). \tag{13.26b-d}$$

By substituting (13.26b-d) in equation (13.25) and separating the real and imaginary parts, we obtain

$$m_0\left(\frac{\partial \mathbf{v}}{\partial t} + \mathbf{v} \cdot \mathbf{v}\right) = -\nabla Q;$$

$$\frac{\partial \mathbf{u}}{\partial t} + \nabla\left(\mathbf{v} \cdot \mathbf{u} + \frac{\lambda^2}{2\tau}\nabla \cdot \mathbf{v}\right) = 0. \tag{13.27a,b}$$

with Q the fractal potential

$$Q = -m_0\frac{\lambda^2}{\tau}\frac{\Delta\sqrt{\rho}}{\sqrt{\rho}} = -\frac{m_0\mathbf{u}^2}{2} - m_0\frac{\lambda^2}{2\tau}\nabla \cdot \mathbf{u}. \tag{13.28}$$

Thus, the fractal potential comes from the non-differentiability of the fractal space-time. Equation (13.27b), by integration up to an arbitrary phase factor which may be set to zero by a suitable choice of the phase ψ, corresponds to the probability conservation law,

$$\partial_t \rho + \nabla \cdot (\rho \mathbf{v}) = 0. \tag{13.29}$$

In the presence of a scalar field U, equation (13.27a) writes

$$m_0 \left(\frac{\partial \mathbf{v}}{\partial t} + \mathbf{v} \cdot \mathbf{v} \right) = -\nabla(Q + U) \tag{13.30}$$

and corresponds to the momentum conservation law. Equations (13.29) and (13.30) stand for the system of equations of hydrodynamics in fractal space-time.

The wave function $\psi(\mathbf{r}, t)$ is invariant when its phase changes by an integer multiple of 2π Indeed, equation (13.26c) gives:

$$\oint m_0 \mathbf{v} \cdot d\mathbf{r} = m_0 \frac{\lambda^2}{\tau} \oint dS = 2\pi n m_0 \frac{\lambda^2}{\tau}; \quad n = 0, \pm 1, \pm 2, \ldots \tag{13.31}$$

as a condition of compatibility between the SR hydrodynamic model and the wave mechanics. Particularly, for $\lambda^2/\tau = \hbar/m_0$, equation (13.31) takes the standard form

$$\oint \mathbf{p} \cdot d\mathbf{r} = nh,$$

i.e. the Ehrenfest relation ([19], [20]).

The set of equations (13.29) and (13.30) represent a complete system of differential equations for the fields $\rho(\mathbf{r}, t)$ and $\mathbf{v}(\mathbf{r}, t)$: relation (13.31) connects each fractal hydrodynamic solution $(\rho, \mathbf{v})_n$ with the wave solution ψ in a unique way.

The field $\rho(\mathbf{r}, t)$ represents a probability distribution density, *i.e.* the probability of finding the particle in the vicinity $d\mathbf{r}$ of the point $\mathbf{r}$ at time t,

$$dP = \rho d\mathbf{r}; \quad \int \int \int \rho \, d\mathbf{r} = 1, \tag{13.32a, b}$$

the integral being extended over the entire domain occupied by the system. Any time variation of the probability density $\rho(\mathbf{r}, t)$ is accompanied by a probability current density $\rho\mathbf{v}$, pointing towards or outwards the corresponding field point $\mathbf{r}$ (equation (13.29)).

The real part of the velocity field $\mathbf{v}$ (equation (13.30)) varies in space and time similarly to a hydrodynamic fluid placed in the force-field of an external potential $U(\mathbf{r}, t)$ and a fractal potential (13.28). The fractal fluid (in the sense of a statistical particles ensemble) exhibits, however, an essential difference as compared to an ordinary fluid: in a rotation motion $\mathbf{v}(\mathbf{r}, t)$ increases (decreases) with the distance from the center $\mathbf{r}$ (equation (13.31)).

The expectation values for the real velocity field and the velocity operator $\hat{\mathbf{v}} = -i(\lambda^2/\tau)\nabla$ of the wave mechanics are equal,

$$\langle \mathbf{v} \rangle = \int\int\int \rho\mathbf{v} d\mathbf{r} = \int\int\int \psi^*\hat{\mathbf{v}}\psi d\mathbf{r} = \langle \hat{\mathbf{v}} \rangle_{WM} \qquad (13.33)$$

but in the higher-order, $|n| > 2$, similar identities are invalid, namely $\langle \mathbf{v}^n \rangle \neq \langle \mathbf{v}^n \rangle_{WM}$. The expectation value for the 'fractal force' vanishes at all times (theorem of Ehrenfest [19], [20]), *i.e.*

$$\langle -\nabla Q \rangle = \int\int\int \rho(-\nabla Q) d\mathbf{r} = 0, \qquad (13.34)$$

since

$$m_0 \frac{\lambda^4}{2\tau^2} \int\int\int \rho\nabla\left(\frac{\nabla^2\sqrt{\rho}}{\sqrt{\rho}}\right) d\mathbf{r} = m_o \frac{\lambda^4}{2\tau^2} \oint (\rho\nabla\nabla \ln\rho)\delta\boldsymbol{\sigma} = 0.$$

$$(13.35)$$

There are two distinctly different types of fractal stationary states:

i) *Fractal dynamic states.* For $\partial/\partial t = 0$ and $\mathbf{v} \neq 0$, equations (13.30) and (13.29) lead to

$$\nabla\left(\frac{1}{2}m_0\mathbf{v}^2 + U + Q\right) = 0; \quad \nabla \cdot (\rho\mathbf{v}) = 0, \qquad (13.36\text{a, b})$$

which yield

$$\frac{1}{2}m_0\mathbf{v}^2 + U + Q = E; \quad \rho\mathbf{v} = \nabla \times \mathbf{F} \qquad (13.37a,b)$$

Consequently, the forces of inertia $m_0\mathbf{v} \cdot \nabla\mathbf{v}$, the external forces $-\nabla U$, and the fractal forces $-\nabla Q$ are in balance at every field point (equation (13.36a)). The sum of the kinetic energy, external potential energy U and fractal potential energy is invariant (equation (13.37a)). Here $E \neq E(\mathbf{r})$ is a constant of integration, and $E \equiv \langle E \rangle$ represents the total energy of the fractal dynamic system. The probability flow density $\rho\mathbf{v}$ has no sources (equation (13.36b)), *i.e.* its streamlines are closed (equation (13.37b)).

ii) *Fractal static states.* For $\partial/\partial t = 0$ and $\mathbf{v} = 0$. equations (13.30) and (13.29) give

$$\nabla(U + Q) = 0, \qquad (13.38)$$

that is

$$U + Q = E. \qquad (13.39)$$

The external force $-\nabla U$ is balanced by the fractal force $-\nabla Q$, at any field point (equation (13.38)). The sum of the external potential energy U, and fractal potential energy Q is invariant, i.e. equal to the integration constant, $E \neq E(\mathbf{r})$. Equation (13.29) is identically satisfied. The quantity $E \equiv \langle E \rangle$ represents the total energy of the fractal static system.

Let us apply the previous formalism to study the atom (fractal atom model). Consider an electron orbiting in the electric field of the nucleus, its potential energy being $U = -Ze^2/(4\pi\varepsilon_0 r)$. The collective chaotic effect on considered electron of all the other electrons has as a result motions on fractal curves. For example, if $\lambda^2/\tau = \hbar/m_0$, a Brownian-like motion can be chosen. A particularization of equations (13.37a,b) for the case of stationary motion gives:

$$\nabla^2 \sqrt{\rho} = -\frac{1}{m_0(\lambda^4/2\tau^2)}\left(E - \frac{1}{2}m_0\mathbf{v}^2 + \frac{Ze^2}{4\pi\varepsilon_0 r}\right)\sqrt{\rho}; \quad \nabla\cdot(\rho\mathbf{v}) = 0;$$

$$0 < r < \infty; \quad 0 \le \varphi \le \pi; \quad 0 \le \Phi \le 2\pi;$$

$$\rho(r=\infty,\varphi,\phi) = 0; \quad \rho(r,\varphi,\phi)_{\varphi\le\pi/2} = \rho(r,\varphi+\pi/2,\phi).$$

$$(13.40a\text{--}g)$$

The probability currents can flow only in closed lines. Because of the azimuthal symmetry of the system, the velocity potential is of the form $S = m\Phi$ (equation (13.26c)), corresponding to a velocity field independent of Φ,

$$\mathbf{v} = \frac{m(\lambda^2/\tau)\mathbf{e}_\phi}{r\sin\varphi}, \quad m = 0, \pm 1, \pm 2, \ldots, \qquad (13.41a,b)$$

where $\mathbf{e}_\phi$ is the unit vector of ϕ direction. This statement is in agreement with the compatibility condition (see equation (13.31)),

$$\oint m_0\mathbf{v}\cdot d\mathbf{r} = mm_0\frac{\lambda^2}{\tau}\oint \frac{\mathbf{e}_\phi\cdot d\mathbf{r}}{r\sin\varphi}$$

$$= mm_0\frac{\lambda^2}{\tau}\int_0^{2\pi} d\Phi = 2\pi n m_0\frac{\lambda^2}{\tau}. \qquad (13.42)$$

Introducing (13.41a) into (13.40a), we have

$$\nabla^2\sqrt{\rho} = -\frac{1}{m_0(\lambda^4/2\tau^2)}\left(E - \frac{m_0m^2(\lambda^4/2\tau^2)}{r^2\sin^2\varphi} + \frac{Ze^2}{4\pi\varepsilon_0 r}\right)\sqrt{\rho}.$$

$$(13.43)$$

Since ρ is independent of ϕ, i.e., $\rho = \rho(r,\varphi)$, $\mathbf{v}\|\mathbf{e}_\phi$, and $\nabla\cdot(\rho\mathbf{v}) = \mathbf{v}\cdot\nabla\rho = 0$ (equation (13.41a))], equation (13.43) can be separated into two differential equations with respect to r and φ, respectively.

Taking

$$\sqrt{\rho} = R(r)\Pi(\varphi), \tag{13.44}$$

we obtain

$$\frac{r^2}{R}\left[\frac{1}{r^2}\frac{d}{dr}\left(r^2\frac{dR}{dr}\right) + \frac{1}{m_0(\lambda^4/2\tau^2)}\left(E + \frac{Ze^2}{4\pi\varepsilon_0 r}\right)R\right] = \overline{\lambda};$$

$$\overline{\lambda} = -\frac{1}{\Pi}\left[\frac{1}{\sin\varphi}\frac{d}{d\varphi}\left(\sin\varphi\frac{d\Pi}{d\varphi}\right) - \frac{m^2}{sin^2\varphi}\Pi\right], \quad \text{(13.45a,b)}$$

where $\overline{\lambda}$ is a separation parameter. Equations (13.45a,b) have solutions only for the eigenvalues of E and $\overline{\lambda}$ constants ([19], [20])

$$\overline{\lambda}_l = l(l+1); \quad l = 0, 1, 2, \ldots, \tag{13.46}$$

and

$$E_n = -\frac{m_0}{2n^2}\left(\frac{Ze^2}{4\pi\varepsilon_0 m_0(\lambda^2/\tau)}\right)^2. \tag{13.47}$$

The solutions of equation (13.45a) are $R(r) \backsim \zeta^l \exp(-\zeta/2) \times L_{n+l}^{2l+1}(\xi)$ ([19], [20]), where $\xi = 2r/na_0$, with

$$u_0 = \frac{4\pi\varepsilon_0 m_0(\lambda^4/\tau^2)}{Ze^2}, \tag{13.48}$$

while the solutions of equation (13.45b) are $\Pi(\varphi) \sim P_l^m(cos\varphi)$. After normalization, one finds ([19], [20]):

$$\rho_{nlm} = C_{nlm}\left(\frac{2r}{na_0}\right)^{2l} e^{-\frac{2r}{na_0}}\left[L_{n+l}^{2l+1}\left(\frac{2r}{na_0}\right)P_l^m(cos\varphi)\right]^2;$$

$$C_{nlm} = \left(\frac{2}{na_0}\right)^3 \frac{(n-l-1)!}{2n[(n+l)!]^3}\frac{2l+1}{4\pi}\frac{(l-|m|)!}{(l+|m|)!}.$$

$$\tag{13.49a,b}$$

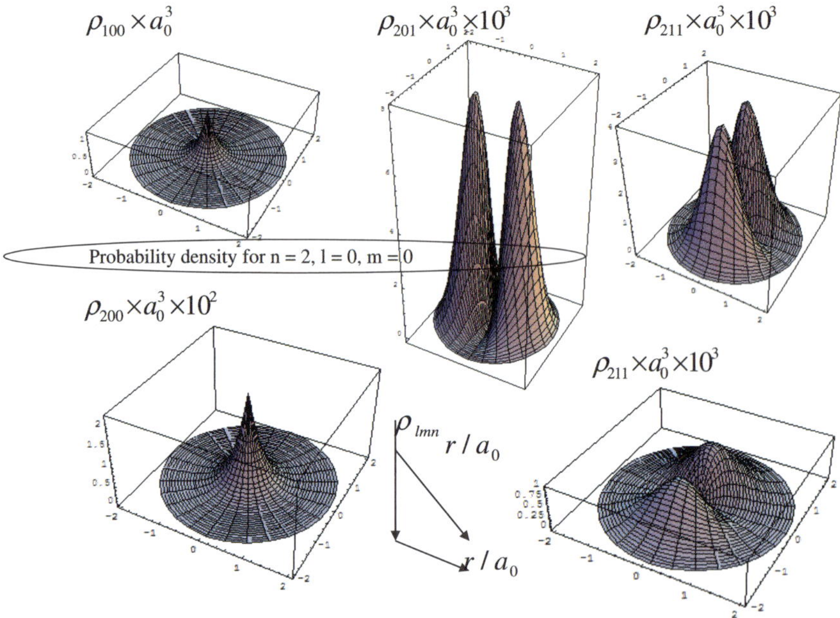

Figure 13.1.   Dependence of the probability density on polar normalized coordinates for various quantum numbers.

For physical reasons, only the following combinations of quantum numbers are acceptable:

$$0 \leq l \leq n - 1; \quad -l \leq m \leq +l. \tag{13.50}$$

Relations (13.41a,b) and (13.49a,b) represent the complete solution $(\rho, \mathbf{v})_{nlm}$ of the SR hydrodynamic model of the fractal atom. In Fig. 13.1 is shown ρ_{nlm} in terms of the polar normalized coordinates for $n = 1, 2$.

By means of the recurrence relations for the associated Laguerre and Legendre polynomials ([19], [20]), one can use the solution (13.49a, b) to show that equation (13.28) becomes

$$Q = -m_0\frac{\lambda^4}{2\tau^2}\left(\frac{1}{n^2 a_0^2} - \frac{2}{a_0 r} + \frac{m^2}{r^2\sin^2\varphi}\right), \tag{13.51}$$

while

$$U(r) = -m_0\frac{\lambda^4}{2\tau^2}\frac{1}{a_0 r}, \tag{13.52}$$

and

$$\frac{1}{2}m_0\mathbf{v}^2 = \frac{m_0 m^2(\lambda^4/2\tau^2)}{r^2\sin^2\varphi}. \tag{13.53}$$

As one can see, equations (13.51)–(13.53) exhibit singularities for $r = 0$, and equations (13.51), (13.53) for $\varphi = 0, \pi$. Since the expectation values of Q, U, and $m_0\mathbf{v}^2/2$ remain finite, these singularities have physical implications. It can be seen that the fractal potential energy Q overcompensates the electrostatic energy U and the kinetic energy $m_0\mathbf{v}^2/2$ at any field point (r, φ, ϕ). The remaining energy is finite and represents the observable energy of the system:

$$\frac{1}{2}m_0\mathbf{v}^2 + U + Q = -m_0\frac{\lambda^4}{2\tau^2}\frac{1}{a_0^2 n^2} \equiv E. \tag{13.54}$$

The states with $m = 0$ are static states ($\mathbf{v} = 0$), while the stated with $m \neq 0$ are dynamic states ($\mathbf{v} \neq 0$) (equations (13.41a,b), (13.51) and (13.53)). In any state with $m \neq 0$, the motion of rotation decreases with increasing distance r, *i.e.* for a given direction φ, $v_\phi \sim 1/r$ (see (13.41a,b)). We note that for $\lambda^2/\tau = \hbar/m_0$, the above presented results are similar to those derived in quantum mechanics ([19], [20]).

13.5. The dispersive approximation of motions in fractal structures. Some properties of the matter

Let us consider the following assumptions: (i) the chaotic effect of the "charge carriers" on the considered "charge" has as result motions on fractal curves with fractal dimension $D_F = 3$; (ii) the ratio dt/τ satisfies the condition $dt/\tau \gg 1$. Then, through the Eq. (13.21), the microparticle movements are described by a generalized Korteweg-de Vries (GKdV) type equation

$$\frac{\partial\mathbf{V}}{\partial t} + \mathbf{V}\cdot\nabla\mathbf{V} + \frac{\lambda^3}{6\tau}\nabla^3\mathbf{V} = 0. \tag{13.55a}$$

This means that in the fractal space-time the dissipative effects can be neglected in comparison with the convection and dispersive ones.

By substituting (13.26b-d) in equation (13.55a), and separating the real and imaginary parts, we obtain the following system:

$$\frac{\partial \mathbf{v}}{\partial t} + \nabla\left(\frac{\mathbf{v}^2}{2} - \frac{\mathbf{u}^2}{2}\right) + \frac{\lambda^3}{6\tau}\nabla^3\mathbf{v} = 0;$$

$$\frac{\partial \mathbf{u}}{\partial t} + \nabla(\mathbf{v}\cdot\mathbf{u}) + \frac{\lambda^3}{6\tau}\nabla^3\mathbf{u} = 0. \tag{13.55b,c}$$

In the one-dimensional differentiable case, for $\mathbf{u} = 0$ or $\rho = const.$, using the dimensionless parameters

$$\overline{\phi} = (v/v_0); \quad \overline{\tau} = \omega_0 t; \quad \overline{\xi} = k_0 X \tag{13.56a--c}$$

and the norming condition

$$\frac{v_0 k_0}{6\omega_0} = \frac{\lambda^3 k_0^3}{6\tau\omega_0} = 1, \tag{13.57}$$

equations (13.55b) take the standard form of (KdV) equation [21]

$$\partial_{\overline{\tau}}\overline{\phi} + 6\overline{\phi}\partial_{\overline{\xi}}\overline{\phi} + \partial_{\overline{\xi}\overline{\xi}\overline{\xi}}\overline{\phi} = 0. \tag{13.58}$$

Through the substitutions

$$w(\theta) = \overline{\phi}(\overline{\xi},\overline{\tau}); \quad \theta = \overline{\xi} - v_f\overline{\tau}, \tag{13.59}$$

equation (13.58), as a result of two successive integrations, becomes

$$\frac{1}{2}w'^2 = F(w) = -\left(w^3 - \frac{v_f}{2}w^2 - gw - h\right), \tag{13.60}$$

where g and h are two constants of integration. If $F(w)$ has real roots, they are of the form

$$e_1 = \overline{w} + 2a\left[\frac{E(s)}{K(s)} - \frac{1}{s^2}\right]; \quad e_2 = \overline{w} + 2a\left[\frac{E(s)}{K(s)} - 1\right];$$

$$e_3 = \overline{w} + 2a\frac{E(s)}{K(s)}, \tag{13.61a--c}$$

with

$$a = \frac{e_3 - e_2}{2}; \quad s^2 = \frac{e_3 - e_2}{e_3 - e_1}; \quad K(s) = \int_0^{\pi/2} (1 - s^2 \sin^2 \varphi)^{-1/2} d\varphi;$$

$$E(s) = \int_0^{\pi/2} (1 - s^2 \sin^2 \varphi)^{1/2} d\varphi. \qquad (13.62\text{a–d})$$

Here $\overline{w}$ is a reference value, and $K(s), E(s)$ the complete elliptic integrals of modulus s [22]. The stationary solution of equation (13.58) then is

$$\overline{\phi}(\overline{\xi}, \overline{\tau}) = \overline{w} + 2a \left(\frac{E(s)}{K(s)} - 1 \right) - 2a \cdot cn^2 \left\{ \frac{\sqrt{a}}{s} \left[\overline{\xi} \right. \right.$$

$$\left. \left. - \left(6\overline{w} + 4a \left(\frac{3E(s)}{K(s)} - \frac{1 + s^2}{s^2} \right) \right) \overline{\tau} + \overline{\xi}_0 \right] ; s \right\} \qquad (13.63)$$

where cn is the Jacobi's elliptic function of modulus s [22], and $\overline{\xi}_0$ an integration constant. As a result, the one-dimensional oscillation modes of the velocity field are of cnoidal type [21] and have the normalized wave length

$$\lambda = 2sK(s)/\sqrt{a} \qquad (13.64)$$

(see Fig. 13.2a–da), the normalized phase velocity

$$v_f = 6\overline{w} + 4a \left[3 \left(\frac{E(s)}{K(s)} \right) - \left(\frac{1 + s^2}{s^2} \right) \right] \qquad (13.65)$$

(see Fig. 13.2a–db), and the normalized group velocity

$$v_g = 6\overline{w} + 4a \left[3\frac{E(s)}{K(s)} - 1 - \frac{1}{s^2} \right.$$

$$+ 3\frac{E^2(s) + 2(s-1)E(s)K(s) - (s-1)K^2(s)}{E(s)K(s) + K^2(s) - sK^2(s)}$$

$$\left. + \frac{2(s-1)K(s)}{s^2[E(s) + K(s) - sK(s)]} \right] \qquad (13.66)$$

(see Fig. 13.2a–dc).

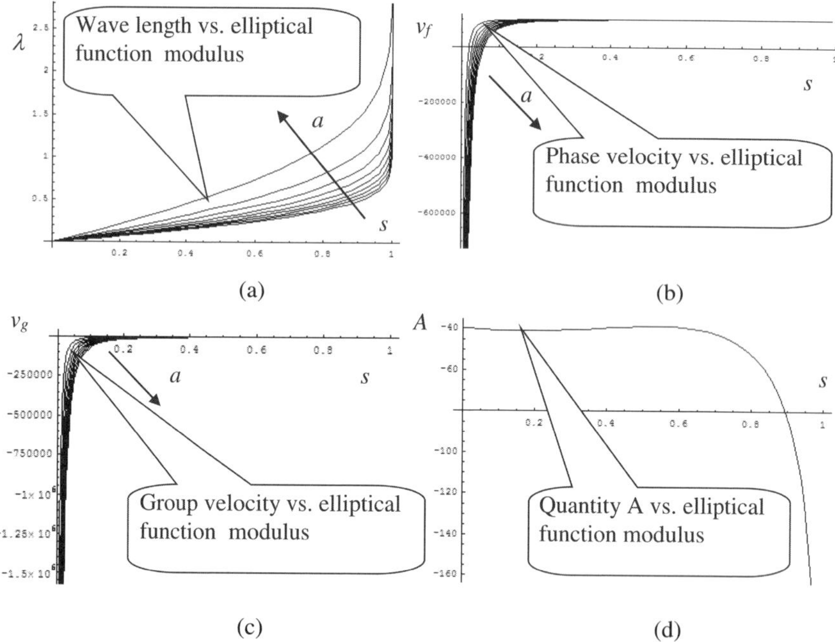

Figure 13.2a–d. Dependencies on s of: (a) The normalized wave length λ; (b) The normalized phase velocity v_f; The group velocity v_g (various values of the parameter a) and (d) of the quantity A.

Let us now discuss the above results:

(i) Through the coefficient $\lambda^3/6\tau$, the parameter s becomes a 'measure' of 'charge' transport type in the considered matter. Thus, the solution (13.63), for $s = 0$, reduces to one-dimensional harmonic wave, and for $s \to 0$, to one-dimensional wave packet. These two subsequences describe the 'charge' transport in a non-quasi-autonomous regime ([21], [23], [24]). For $s = 1$, solution (13.63) becomes an one-dimensional soliton [21], while for $s \to 1$, one-dimensional soliton packet. These last two subsequences describe 'charge' transport in a quasi-autonomous regime ([21], [23], [24]).

(ii) By eliminating the parameter a from relations (13.64) and (13.65), one obtains

$$(v_f - 6\overline{w})\lambda^2 = A(s);$$

$$A(s) = 16\left[3s^2 E(s)K(s) - (1 + s^2)K^2(s)\right],$$

(13.67a,b)

where the quantity $A(s)$ is plotted in Fig. 13.2a–d. We observe that for $s = 0 : 0.7$, $A(s) = const.$, and consequently equation (13.67a) takes the form $(v_f - 6\overline{w})\lambda^2 = const.$ Therefore, in the differentiable case, the 'charge' transport is controlled through the flowing regimes of the fractal fluid, and the separation between them is given by the 0.7 value of the parameter s.

(iii) The previous results show through the normalized group velocity (13.66) an increase of the 'charge' transport by means of quasi-autonomous structures. This theoretical result explains some 'anomalies' that were experimentally observed in nanostructures, e.g. the increase of the thermal conductance in nanofluids ([25]–[27]), etc.

Let us now investigate the previous phenomenon in the nondifferential case. This can be achieved by the substitutions $\phi = (v_f/4)f^2$ and $i\eta = (v_f/4)^{1/2}\theta$ in equation (13.60). Moreover, this equation with $h = 0$, becomes

$$\partial_{\eta\eta}f = f^3 - f,$$

i.e. a Ginzburg-Landau (GL) type equation ([21], [28]). These results show that: (i) The coordinate η has a dynamic significance, and the variable f a probabilistic significance, while the space-time becomes fractal (for details see [10]–[14]). (ii) Since the general solution of GL equation can be expressed, with adequate normalization and choice of the integration constants, by means of the elliptic function $f(\eta) = sn(\eta; s)$ (for details see [29]), the 'charge' transport is controlled by the fractal potential [30]

$$Q = -\frac{1}{f}\frac{d^2 f}{d\eta^2} = (1 - f^2) = cn^2(\eta, s), \qquad (13.68)$$

also through cnoidal oscillation modes. Thus, as in the previous differentiable case, $(s = 0, s \to 0)$ implies a non-quasi-autonomous regime, while $(s = 1, s \to 1)$ implies the quasi-autonomous regime. (iii) For $s = 1$ the general solution of GL equation is the fractal kink $f_k(\eta) = \tanh(\eta)$. The fractal kink spontaneously breaks the "vacuum" symmetry by tunneling, and generates coherent structures ([29], [30]). This mechanism is similar to the one of superconductivity and can explain the properties of superconductors through the Cooper pairs [29]. (iv) The expression of the normalized fractal potential takes a very simple form, which is directly proportional to the density of states of the fractal fluid — see equation (13.68). When the density of states f^2 becomes zero *i.e.* in the absence of the vacuum symmetry spontaneous breaking, the fractal potential takes a finite value, $Q = 1$. The fractal fluid is normal (it works in a non-quasi-autonomous regime) and there are no coherent structures in it. If f^2 becomes 1, i.e. in the presence of the vacuum symmetry spontaneous breaking, the fractal potential equals zero, i.e. the entire quantity of energy of the fractal fluid is transferred to its coherent structures. Then the fractal fluid becomes coherent (it works in a quasi-autonomous regime). Therefore, one can assume that the energy from the fractal fluid can be stocked by transforming all the environments entities into coherent structures and then 'freezing' them. The fractal fluid acts as an energy accumulator through the fractal potential (13.68). (v) The correlation between the differentiable and the non-differentiable scales implies the equivalence theorem in periods of the elliptic functions (13.68) and (13.63) [20]. Then, the fractal space-time is of Cantor-type (for details see [29]–[31]). Moreover, the 'charge' transport implies at any scale a fractal. Such a result is obtained through the iterative map induced by the elliptic function cn^2 for various values of the parameter s — Figures 13.3a–fa–f. For any object given in these drawings, the Hausdorff-Besicovitch theorem is respected [15].

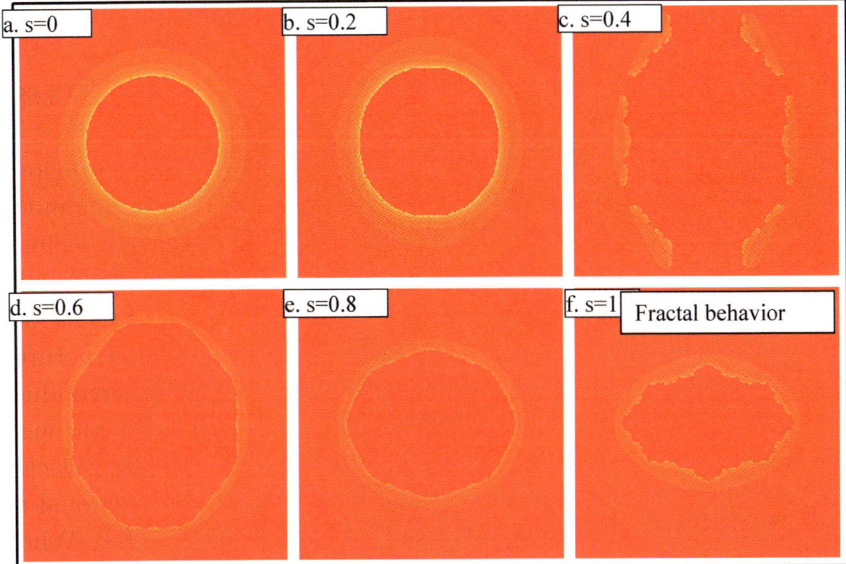

Figure 13.3a–f. The iterative map induced by the elliptic function cn^2 for different values of the parameter s.

Here are some concluding remarks:

(i) Using the mathematical formalism of the scale relativity theory, an extension of the model for an arbitrary fractal dimension is established.

(ii) It follows, through an equation of motion for the complex speed field, that in a fractal fluid the convection, dissipation and dispersion are reciprocally compensating at any scale (differentiable or non-differentiable). From here, for an irrotational movement, a generalized Schrödinger equation is obtained.

(iii) For $D_F = 2$ and dissipative approximation of motions in the fractal structures, *i.e.* in the absence of dispersion, a generalized Navier-Stokes type equation results. Applying this model to the study the atom, it follows that the real part of the complex velocity field describes the electron averaged movement. In this case, the electron moves on stationary orbits according to quantification conditions. The imaginary part of the complex velocity describes the fractality through a fractal potential. Now, using this potential, from the averaged movements (on stationary orbits), the electron energy quantization also follows. If the viscosity coefficient is taken as $(\lambda^2/\tau) = \hbar/m_0$, the classical results of quantum mechanics for the hydrogen atom are obtained.

(iv) For $D_F = 3$ and dispersive approximation of motions in the fractal structures, *i.e.* in the absence of dissipation, a generalized Korteweg de Vries type equation is obtained. At differentiable scale, the 'charge' transport is achieved by one-dimensional cnoidal oscillation modes of the velocity field. For different degrees of 'charge' coupling, the one-dimensional cnoidal velocity oscillation modes contain one-dimensional velocity harmonic waves, one-dimensional velocity wave packet, one dimensional velocity soliton packet, and one dimensional velocity soliton. The first two subsequences describe the non-autonomous regime of the 'charge' transport, while the last ones describe the quasi-autonomous regime of the 'charge' transport. In the non-autonomous regime, a relation between the normalized wave length and the normalized phase velocity, *i.e.* a dispersion type relation, is obtained. These two regimes are separated by "0.7 structure" that is experimentally observed for various nanostructures [32]. At non-differentiable scale, the fractal potential acts as an energy accumulator and controls through the coherence the 'charge' transport. In this context, it follows that the superconductivity is exclusively a fractal property.

(v) The correlation between differentiable and the non-differentiable scales implies a fractal space-time of Cantor type. Then, the 'charge' transport implies, at any scale, a fractal.

(vi) These results extend on both the old hydrodynamic model of quantum mechanics [33], and on the more recent chaotic-stochastic model of the atom ([34], [35]), as well.

References

[1] Schrödinger E. 1928 *Collected Papers on Wave Mechanics* (London: W. M. Deans).

[2] Green H. S. 1965 *Matrix Mechanics* (Groningen: P. Noordhoff Ltd.).

[3] Feynman R. P. and Hibbs A. R. 1965 *Quantum Mechanics and Path Integrals* (New York: McGraw-Hill).

[4] Madelung E., *Z. Physik* 1926 **40** 322.

[5] De Broglie L., 1927 Compt. Rend. **185** 380.

[6] Takabayasi T., 1952 Progr. Theor. Phys. (Kyoto) **8** 143.

[7] Bohm D., 1952 Phys. Rev. **85** 166.

[8] De Broglie L., 1967 Compt. Rend. **264** 1041.

[9] Bohm D. and Bub J., 1966 Rev. Mod. Phys. **38**38 453.

[10] Nottale L., 1993 *Fractal Space-Time and Microphysics: Towards a Theory of Scale Relativity* (Singapore: World Scientific); Nottale L., 1997, *Astron. Astrophys.*, **327**, 867; Nottale L., 1998, *Chaos, Solitons and Fractals*, **9**, 1051; Nottale L., 1999, *Chaos, Solitons and Fractals*, **10**, 459; Nottale L., 2003, *Chaos, Solitons and Fractals*, **16**, 539; Da Rocha D., and Nottale L. 2003, *Chaos, Solitons and Fractals*, **16**, 565; Nottale L., 2005, *Chaos, Solitons and Fractals*, **25**, 797.

[11] Nottale L. 1992 *Int. J. Mod. Phys.* **A7** 4899.

[12] Nottale L. and Célérier M. N. 2007 *J. Phys. A: Math. Theor.* **40** 14471.

[13] Ord G. N. 1983 *J. Phys. A* **16** 1869.

[14] Weiberl P., Ord G., Rössler O. E., 1005 *Space Time Physics and Fractality*, New York, Springer.

[15] Mandelbrot B. 1982 *The Fractal Geometry of Nature* (San Francisco: Freeman).

[16] Nottale L., Célérier M. N. and Lehner T. 2006 *J. Math. Phys.* **47** 032303.

[17] Célérier M. N. and Nottale L. 2004 *J. Phys. A: Math. Gen.* **37** 931.

[18] Gottlieb I., Agop M., Ciobanu G., and Stroe A. 2006 (*Chaos, Solitons and Fractals* **30** 380.

[19] Phillips A. C. 2003 *Introduction to Quantum Mechanics* (New York: John Wiley and Sons Ltd).

[20] Ballentine L. E. 1998 *Quantum mechanics. A Modern Development* (Singapore: World Scientific).

[21] Jackson E. A. 1991 *Perspectives in Nonlinear Dynamics*, vol. I+II (Cambridge: Cambridge University Press).

[22] Bowman F., 1955 *Introduction to Elliptic Function with Applications* (London: English University Press).

[23] Imry Y. 2002 *Introduction to Mesoscopic Physics* (Oxford: Oxford University Press).

[24] Ferry D. K. and Goodnick S. M. 1997 *Transport in Nanostructures* (Cambridge: Cambridge University Press).

[25] Jang S. P. and Choi S. U. S. 2004 *Appl. Phys. Lett.* **84** 4316.

[26] Kumar D. H., Patel H. E., Kumar V. R. R., Sundararajan T., Pradeep T., and Das S., K. 2004 *Phys. Rev. Lett.* **93** 144301.

[27] Prasher R, Bhattacharya P. and Phelan P. E. 2005 *Phys Rev Lett.* **94**, 025901.

[28] Poole C. P., Farach H. A. and Creswick R. J., 1995 *Superconductivity* (San Diego-New York-Boston-London-Sydney-Tokyo-Toronto: Academic Press).

[29] Agop M., Ioannou P. D. and Nica P. 2005 *J. Math. Phys.* **46** 062110.

[30] Agop M., Nica P. and Girtu M. 2008 *Gen. Relativ. Gravit.* **40** 35–55.

[31] Agop M., Nica P., Ioannou P. D., Malandraki O. and Gavanas-Pahomi I. 2007 *Chaos, Solitons and Fractals* **34** 1704.

[32] Chatti O., Nicholls J. T., Proskuryakov Y. Y., Lumpkin N., Farrer I. and Ritchie D. A. 2006 *Phys. Rev. Lett.* **97** 056601.

[33] Wilhem H. E. 1970 *Phys. Rev. D* **1** 2278.

[34] Argyris J. and Ciubotariu C. 2000 *Chaos, Solitons and Fractals* **11** 1001.

[35] Ciubotariu C., Stancu V. and Ciubotariu C. C. 2002 *Fundamental Theories of Physics* **126** 357.

Chapter 14

Extended Fractal Hydrodynamic Model with an Arbitrary Fractal Dimension and its Implications

14.1. Introduction

The complex dynamical systems which display chaotic behavior are recognized to acquire self-similarity and manifest strong fluctuations at all possible scales (for details see [1]–[9]). Since the fractality appears as an universal property of the complex systems, it is necessary to construct a new physics (fractal physics), either through the non-differentiability (for mathematical procedures see [10], [11] and for physical implications see ([5], [6], [7], [12]) or through the fractional derivatives ([13], [14]). Thus, both the scale relativity (SR) ([5], [6], [15]–[27]) and the fractional physics ([13], [14]) were developed. In the following investigation we shall refer only to the SR model (Nottale's model [5]–[7]), *i.e.* to a continuous but non-differentiable physical model. The SR uses the word "non-differentiability" in its general meaning ([5], [6], [10], [11], [15]–[27]), denoting a set that shows structures at all scales, this way being explicitly resolution-dependent. More precisely, one can demonstrate ([5], [6], [10], [11]) that the measure of a continuous, almost everywhere non-differentiable set of topological dimension D_T is a function of resolution, $L = L(\varepsilon)$, and diverges when resolution tends to zero, $L(\varepsilon \to 0) \to \infty$. In such a framework, resolutions are considered to be inherent to the description of the new, fractal space-time.

SR involves a scale dependence on the reference frames. Therefore, it is possible to add to the standard variables (position, orientation and motion), which characterize the reference frame, other variables characterizing its scale state. The use of differential equations is made possible due to the representation of physical quantities, usually mere functions of the space-time coordinates $f(x)$, by fractal functions, $f[x(\varepsilon), \varepsilon]$, explicitly depending on the scale variables, generically denoted by ε.

Generalizing the definition of fractal functions to fractal space-times (fractal space-time $\equiv$ equivalence class of a family of Riemannian space-times), we obtain scale dependent geodesics equations and, therefore, an infinite family of geodesics. In the SR we have some fundamental conditions, as consequences of the non-differentiability ([5], [6], [15]–[27]). These consequences have been exposed in the previous chapter.

Note that, so far. the following levels of the SR have been considered:

(i) a 'Galilean' version corresponding to the standard fractals with constant fractal dimensions, the dilatation laws being the usual ones. This theory provides us a new foundation of quantum mechanics from the first principles ([5], [6], [21], [22]);

(ii) a special scale-relativistic version that implements, in a more general way, the principle of scale relativity. It yields new dilatation laws of a Lorentzian form that demand to re-interpret the Planck length scale as a lower, impassable scale, invariant under dilatation. The predictions of such a theory depart from that of standard quantum mechanics at large energies ([5], [6], [16]–[20]);

(iii) the third level, 'general scale-relativistic' version of the theory deals with non-linear scale laws and accounts for the coupling between scale laws and motion laws. It yields a new interpretation of gauge invariance and allows one to get new mass-charge relations that solve the scale-hierarchy problem ([5], [15]).

Using this theory, both conceptual (the complex nature of wave function, probabilistic nature of quantum theory, principle of correspondence, quantum-classical transition, divergence of masses and charges, nature of Planck scale, nature and quantization of electric charge, origin of mass discretization of elementary particles, nature of cosmological constant, etc.) and quantized (the mass-charge relations, electro-weak scale, electron scale, elementary fermion mass spectrum, quantization of planetary motion, red-shift quantization in binary galaxies, etc.) results are obtained ([5], [6], [15]–[27]).

In the SR context, the fractal dimension $D_F = 2$ plays a very important role. It represents the dimension in which, for the Markov-Wiener processes ([5], [6]) and $D = \hbar/2m_0$ (where $\hbar$ is the reduced Planck's constant and m_0 the rest mass of the test particle) the SR model reduces itself to the standard Schrödinger equation and, thus, to the study of the quantum system dynamics. Moreover, for $D = GM_0/2w_0$ and the Newtonian potential $U = GM_0m_0/r$ (with G Newton's constant, M_0 the rest mass of the source field, $w_0 \approx 144km/s$ [28], and r the relative distance between the source field and the test particle), we obtain a Schrödinger-type equation and, thus, the quantization of planetary motion at infragalactic scale ([5], [27]), or the red-shift quantization in binary galaxies at extragalactic scale ([5], [27]). Therefore, according to the above-mentioned considerations, the SR model can be used both in studying microscopic scale phenomena (subatomic dynamics, quantum systems dynamics, etc.) and macroscopic ones (*e.g.* atmospheric and stellar plasma physics).

Since the theoretical description of microphysical systems is based on Schrödinger's wave mechanics [29], on Feynmans path-integral mechanics [30], or on the hydrodynamic model [31], we expect that, scale-independently, *i.e.* similar to the SR model (in any of its versions: hydrodynamic, Schrödinger type, etc.), the non-differentiability can be rephrased starting from functional analysis, fractional action-like or wavelets analysis. This is now possible, as it follows from Refs. ([10]–[12], [32]–[35]).

In the present chapter, using the RT in its hydrodynamic version for an arbitrary constant fractal dimension, some applications are obtained by considering the motions of particles on fractal curves of constant fractal dimension D_F and second order terms in the motion equation of the complex velocity field.

14.2. Fractal hydrodynamic model for an arbitrary constant fractal dimension

Let us suppose that the motion of the particles takes place on continuous but non-differentiable curves, *i.e.* on fractals ([1], [3], [4]). The "non-differentiability" in an arbitrary constant fractal dimension D_F implies replacement of the standard time derivative d/dt by a new complex operator $\hat{\partial}/\partial t$ (for details, see the complex operator (18) in Ref. [25]),

$$\frac{\hat{\partial}}{\partial t} = \frac{\partial}{\partial t} + \mathbf{V} \cdot \nabla - i\frac{\lambda^2}{2\tau}\left(\frac{dt}{\tau}\right)^{\left(\frac{2}{D_F}\right)-1}\Delta, \qquad (14.1)$$

where $\mathbf{V}$ is the complex velocity field, λ — a characteristic scale length. and τ a characteristic scale time. Consequently, the covariant form of Newtons first law in the fractal space-time writes

$$\frac{\hat{\partial}\mathbf{V}}{\partial t} = \frac{\partial\mathbf{V}}{\partial t} + \mathbf{V} \cdot \nabla\mathbf{V} - i\frac{\lambda^2}{2\tau}\left(\frac{dt}{\tau}\right)^{\left(\frac{2}{D_F}\right)-1}\Delta\mathbf{V} = 0. \qquad (14.2)$$

If the motion of the fractal fluid is irrotational, $\mathbf{\Omega} = \nabla \times \mathbf{V} = 0$, we can choose $\mathbf{V}$ of the form

$$\mathbf{V} = -i\frac{\lambda^2}{\tau}\left(\frac{dt}{\tau}\right)^{\left(\frac{2}{D_F}\right)-1}\nabla \ln \Psi, \qquad (14.3)$$

where Ψ is a scalar complex function. Then, equation (14.2) becomes a Navier-Stokes type equation,

$$\frac{\hat{\partial}\mathbf{V}}{\partial t} = \frac{\partial\mathbf{V}}{\partial t} + \nabla\left(\frac{\mathbf{V}^2}{2}\right) - i\frac{\lambda^2}{2\tau}\left(\frac{dt}{\tau}\right)^{\left(\frac{2}{D_F}\right)-1}\Delta\mathbf{V} = 0, \qquad (14.4)$$

with an imaginary coefficient of viscosity, ν

$$\nu = i\frac{\lambda^2}{2\tau}\left(\frac{dt}{\tau}\right)^{\left(\frac{2}{D_F}\right)-1}, \qquad (14.5)$$

while in terms of function Ψ, a Schrödinger type equation is obtained

$$\frac{\lambda^4}{4\tau^2}\left(\frac{dt}{\tau}\right)^{\left(\frac{4}{D_F}\right)-2}\Delta\Psi + i\frac{\lambda^2}{2\tau}\left(\frac{dt}{\tau}\right)^{\left(\frac{2}{D_F}\right)-1}\frac{\partial\Psi}{\partial t} = 0. \qquad (14.6)$$

The presence of an imaginary viscosity coefficient shows the following facts: (i) at macroscopic scale, the behavior of the fractal fluids is of viscoelastic type or hysteretic type. Such a result is in agreement with the ideas developed in [36], [37]: the fractal fluid can be described by Kelvin-Voight or Maxwell rheological model with complex structure coefficients [36] (particularly, with the imaginary viscosity coefficient (14.5)). Thus, such materials are endowed with "memory"; (ii) at microscopic scale, the scalar field of the complex velocity has a stochastic behavior. In particular, at Compton scale $(D = \lambda^2/2\tau = \hbar/2m_0)$ and motions on Peano curves ([5], [6]), *i.e.* for the fractal dimension $D_F = 2$, the scalar field of the complex velocity is also a wave function, so that equation (14.6) takes the form of the "standard" Schrödinger equation:

$$\frac{\hbar^2}{2m_0}\Delta\Psi + i\hbar\frac{\partial\Psi}{\partial t} = 0.$$

This means that the Schrödinger equation emerges from a Navier-Stokes type equation with an imaginary viscosity coefficient $\nu = \frac{i\hbar}{2m_0}$, for irrotational motion of a fractal fluid on Peano curves $(D_F = 2)$ at Compton scale.

Let us now choose function Ψ in the form $\Psi = \sqrt{\rho}e^{iS}$, with $\sqrt{\rho}$ the amplitude, and S the phase. In this case, the complex velocity field (14.3)

$$\mathbf{V} = \mathbf{v} - i\mathbf{u} \qquad (14.7)$$

has the components

$$\mathbf{v} = \frac{\lambda^2}{\tau}\left(\frac{dt}{\tau}\right)^{\left(\frac{2}{D_F}\right)-1}\nabla S; \quad \mathbf{u} = \frac{\lambda^2}{2\tau}\left(\frac{dt}{\tau}\right)^{\left(\frac{2}{D_F}\right)-1}\nabla\ln\rho,$$

$$(14.8\text{a, b})$$

where $\mathbf{v}$ is the real (differentiable) part, and $\mathbf{u}$ the imaginary (non-differentiable, or fractal) part of the velocity vector field.

Equations (14.8a,b) which define the components of the complex velocity field of the fractal fluid are more general than those obtained from Nottale's SR model ([5], [6], [15]–[22]). For the fractal dimension $D_F = 2$, Nottale's results are refound ([5], [6], [15]–[22]):

$$\mathbf{v} = \frac{\lambda^2}{\tau} \nabla S; \quad \mathbf{u} = \frac{\lambda^2}{2\tau} \nabla \ln \rho.$$

Introducing (14.7) and (14.8a,b) in (14.4) and separating the real and imaginary parts, *i.e.* through separation of the movements at differentiable scale from those at non-differentiable scale, we obtain:

$$\frac{\partial \mathbf{v}}{\partial t} + \nabla \left(\frac{\mathbf{v}^2}{2} - \frac{\mathbf{u}^2}{2} - \frac{\lambda^2}{2\tau} \left(\frac{dt}{\tau} \right)^{\left(\frac{2}{D_F}\right)-1} \nabla \cdot \mathbf{u} \right) = 0;$$

$$\tag{14.9}$$

$$\frac{\partial \mathbf{u}}{\partial t} + \nabla \left(\mathbf{v} \cdot \mathbf{u} + \frac{\lambda^2}{2\tau} \left(\frac{dt}{\tau} \right)^{\left(\frac{2}{D_F}\right)-1} \nabla \cdot \mathbf{v} \right) = 0,$$

or, up to an arbitrary phase factor which can be set to zero by a suitable choice of the phase of Ψ,

$$m_0 \left(\frac{\partial \mathbf{v}}{\partial t} + \mathbf{v} \cdot \nabla \mathbf{v} \right) = -\nabla Q;$$

$$\tag{14.10a,b}$$

$$\frac{\partial \rho}{\partial t} + \nabla \cdot (\rho \mathbf{v}) = 0,$$

with Q the fractal potential

$$Q = -m_0 \frac{\lambda^4}{2\tau^2} \left(\frac{dt}{\tau} \right)^{\left(\frac{4}{D_F}\right)-2} \frac{\Delta \sqrt{\rho}}{\sqrt{\rho}}$$

$$= -\frac{m_0 \mathbf{u}^2}{2} - m_0 \frac{\lambda^2}{2\tau} \left(\frac{dt}{\tau} \right)^{\left(\frac{2}{D_F}\right)-1} \nabla \cdot \mathbf{u}. \tag{14.11}$$

As it is seen, the fractal potential depends only on the imaginary part $\mathbf{u}$ of the complex velocity field $\mathbf{V}$, and comes from the non-differentiability of the fractal space-time. Equation (14.10a), *i.e.* the momentum conservation law, together with equation (14.10b), *i.e.* the probability conservation law, stand for the basis of fractal hydrodynamic model.

In such a representation, the presence of an electromagnetic field allows one to develop of a fractal magneto-hydrodynamic (MHD) model. To this end, we shall first replace the standard derivative by the covariant derivative in equation (14.6)

$$\nabla_i = \partial_i - \frac{e}{m_0}\frac{\tau}{\lambda^2}\left(\frac{dt}{\tau}\right)^{1-\left(\frac{2}{D_F}\right)} A_i,$$

where A_i is the vector potential of the electromagnetic field, then introduce the wave function $\Psi = \sqrt{\rho}e^{iS}$ into the resulting equation, and finally separate the real and imaginary parts using relations (14.8a,b).

Two types of stationary states can be distinguished:

i) *Dynamic states.* For $\partial/\partial t = 0$ and $\mathbf{v} \neq 0$, *i.e.* at the differentiable scale. Equations (14.10a,b) yield

$$\nabla\left(m_0\frac{\mathbf{v}^2}{2} - m_0\frac{\mathbf{u}^2}{2} - m_0\frac{\lambda^2}{2\tau}\left(\frac{dt}{\tau}\right)^{\left(\frac{2}{D_F}\right)-1}\nabla\cdot\mathbf{u}\right) = 0;$$

$$(14.12a,b)$$

$$\nabla\cdot(\rho\mathbf{v}) = 0,$$

and, by integration:

$$m_0\frac{\mathbf{v}^2}{2} - m_0\frac{\mathbf{u}^2}{2} - m_0\frac{\lambda^2}{2\tau}\left(\frac{dt}{\tau}\right)^{\left(\frac{2}{D_F}\right)-1}\nabla\cdot\mathbf{u} = E;$$

$$(14.13a,b)$$

$$\rho\mathbf{v} = \nabla\times\mathbf{F}.$$

Consequently, the non-fractal force of inertia $m_0\mathbf{v}\cdot\nabla\mathbf{v}$ and the fractal force $-\nabla Q$ are in balance at every point —

equation (14.12a). The sum of the non-fractal kinetic energy, $m_0\mathbf{v}^2/2$ and the fractal potential Q is invariant, *i.e.* equal to the integration constant $E \neq E(\mathbf{r})$ — equation (14.13a). In its turn, $E \equiv \langle E \rangle$ represents the total energy of the dynamical system. The probability flow $\rho\mathbf{v}$ has no sources — equation (13.12b), *i.e.* its streamlines are closed — equation (14.13b). If an external potential U has to be considered, equation (14.13a) becomes

$$m_0\frac{\mathbf{v}^2}{2} - m_0\frac{\mathbf{u}^2}{2} - m_0\frac{\lambda^2}{2\tau}\left(\frac{dt}{\tau}\right)^{\left(\frac{2}{D_F}\right)-1}\nabla\cdot\mathbf{u} + U = E. \qquad (14.13c)$$

ii) *Static states.* For $\partial/\partial t = 0$ and $\mathbf{v} = 0$, that is at the non-differentiable scale, equations (14.10a,b) give

$$\nabla\left(-m_0\frac{\mathbf{u}^2}{2} - m_0\frac{\lambda^2}{2\tau}\left(\frac{dt}{\tau}\right)^{\left(\frac{2}{D_F}\right)-1}\nabla\cdot\mathbf{u}\right) = 0, \qquad (14.14a)$$

which yields

$$-m_0\frac{\mathbf{u}^2}{2} - m_0\frac{\lambda^2}{2\tau}\left(\frac{dt}{\tau}\right)^{\left(\frac{2}{D_F}\right)-1}\nabla\cdot\mathbf{u} = E. \qquad (14.14b)$$

Thus, the fractal force $-\nabla Q$ has a value of zero — equation (14a). The fractal potential Q is invariant, *i.e.* equal to the constant of integration $E \neq E(\mathbf{r})$ — equation (14.14b). Here $E \equiv \langle E \rangle$ represents the total energy of the static system. Equation (14.10b) is identically satisfied.

If an external potential U is also considered, equation (14.14b) writes

$$-m_0\frac{\mathbf{u}^2}{2} - m_0\frac{\lambda^2}{2\tau}\left(\frac{dt}{\tau}\right)^{\left(\frac{2}{D_F}\right)-1}\nabla\cdot\mathbf{u} + U = E. \qquad (14.14c)$$

As illustrations of the fractal hydrodynamic formalism, several static and time-dependent fractal systems are further analyzed.

14.3. Particle in a box in a fractal space-time

One of the simplest fractal systems is a particle enclosed in a one-dimensional box of infinite potential walls at $x = 0$ and $x = a$

$$U(x) = 0,\ 0 < x < a; \quad U(x \le 0) = \infty;\ U(x \ge a) = \infty.$$
$$(14.15\text{a--d})$$

In the stationary state

$$\rho(x)\,\mathbf{v}(x) = \mathbf{c}\ (const.) \tag{14.16}$$

according to equation (14.12b), and either one of the boundary conditions

$$\rho(0) = 0; \quad \rho(a) = 0 \tag{14.17a, b}$$

shows that $\mathbf{c} = 0$. Hence

$$\mathbf{v}(x) = 0. \tag{14.18}$$

Equation (14.14c) then applies accordingly, leading (under consideration of equations (14.15a–d)) to the linear boundary value problem

$$\frac{d^2\sqrt{\rho}}{dx^2} + \frac{E}{2m_0 D^2}\sqrt{\rho} = 0; \quad D = \frac{\lambda^2}{2\tau}\left(\frac{dt}{\tau}\right)^{\left(\frac{2}{D_F}\right)-1};$$

$$0 < x < a; \quad \rho(0) = \rho(a) = 0. \tag{14.19a--d}$$

A solution $\rho = \rho_n$ exists if the total energy of the system has certain eigenvalues $E = E_n$. One finds

$$\rho_n = \frac{2}{a}\sin^2\left(\frac{n\pi}{a}x\right) \tag{14.20}$$

and

$$E_n = 2m_0 D^2 \left(\frac{n\pi}{a}\right)^2; \quad n = 0, 1, 2\ldots \tag{14.21}$$

In particular, for $D_F = 2$ and $D = \hbar^2/2m_0$, we are left with the results of quantum mechanics ([38], [39]):

$$\Psi_n(x) = \sqrt{\frac{2}{a}} \sin\left(\frac{n\pi}{a}x\right); \quad E_n = \frac{\hbar^2}{2m_0}\left(\frac{n\pi}{a}\right)^2, \quad n = 0, 1, 2\ldots$$

where Ψ_n are the normalized wave functions of the stationary states and E_n the discrete eigenvalues of the energy corresponding to these states.

According to our model, equations (14.18) and (14.20) have the following interpretations:

i) At the differentiable scale, the real (differentiable) part of the complex velocity field of the fractal fluid, $\mathbf{v}$, is zero — see equation (14.18). According to equation (14.8a), the fractal fluid is coherent, *i.e.* the fluid particles have the same phase. This kind of fluid has a superconductor or super-fluid type behavior.

ii) At the non-differentiable scale, the imaginary (non-differentiable or fractal) part of the complex velocity field of the fractal fluid, $\mathbf{u}$, is non-zero, and its x-component is (Equation (14.8b))

$$u_x = D\frac{d}{dx}(\ln \rho) = 2D\frac{n\pi}{a}\cot\left(\frac{n\pi}{a}x\right). \tag{14.22}$$

Then, the fractal one-dimensional potential (14.11) becomes

$$Q_n = -\frac{m_0 u_x^2}{2} - m_0 D\frac{du_x}{dt}$$

$$= -2m_0 D^2 \left(\frac{n\pi}{a}\right)^2 \cot^2\left(\frac{n\pi}{a}x\right) + 2m_0 D^2 \left(\frac{n\pi}{a}\right)^2 \frac{1}{\sin^2\left(\frac{n\pi}{a}x\right)}$$

$$= 2m_0 D^2 \left(\frac{n\pi}{a}\right)^2. \tag{14.23}$$

The identity of relations (14.21) and (14.23) shows that the "observable" in the form of energy generated by the fractal component of the complex velocity field is quantized by means of the fractal potential (14.23).

Therefore, in our model the particle in box in a fractal space-time is equivalent to a coherent fractal fluid whose particles move on stationary trajectories which satisfy condition (14.23). The complex velocity field of the fractal fluid proves to be essential: the zero value of the real (differential) part specifies the coherence of the fractal fluid, while the non-zero value of the imaginary (fractal) part implies through condition (14.23) the stationary trajectories of the fluid particles. The momentum transfer is achieved only through the fractal component of the complex velocity field.

It also follows a new physical meaning associated with the concept of particle: various properties of "particles" can be reduced to the geometric structures of the (non-differentiable) geodesics of the fractal space-time (such a subject is also analyzed in [5], [6], [22]–[27]).

In plasma, the coherent fractal fluids can be associated with double layers-type or multiple double layers-type patterns ([40], [41]) and condition (14.23) with a "quantization" of the potentials which generate the multiple double layers (for details see [40], [41]).

14.4. Harmonic oscillator in a fractal space-time

The one-dimensional harmonic oscillator in a fractal space-time is a particle in a spatially symmetric field whose potential energy has the form

$$U(x) = \frac{1}{2} m_0 \Omega^2 x^2, \quad -\infty \le x \le +\infty. \tag{14.24a, b}$$

According to equation (14.12b), in the stationary state

$$\rho(x)\mathbf{v}(x) = \mathbf{c}(const.). \tag{14.25}$$

Since either one of the boundary conditions gives

$$\rho(-\infty) = 0; \quad \rho(+\infty) = 0, \tag{14.26a, b}$$

we have $\mathbf{c} = 0$. Hence

$$\mathbf{v}(x) = 0. \tag{14.27}$$

Then equation (14.14c), in view of considerations accompanying equations (14.24a,b), leads to the linear boundary-value problem

$$\frac{d^2\sqrt{\rho}}{dx^2} + \left[\left(\frac{E}{2m_0 D^2}\right) - \left(\frac{\Omega}{2D}\right)^2 x^2\right]\sqrt{\rho} = 0;$$

$$-\infty \leq x \leq +\infty; \quad \rho(x = -\infty) = \rho(x = +\infty) = 0. \qquad (14.28\text{a}-\text{c})$$

By introducing the dimensionless variable

$$\xi = \sqrt{\frac{\Omega}{2D}}x, \qquad (14.29)$$

equations (14.28a–c) take the form

$$\frac{d^2\sqrt{\rho}}{d\xi^2} + \left(\frac{E}{m_0 D\Omega} - \xi^2\right)\sqrt{\rho} = 0;$$

$$-\infty \leq \xi \leq +\infty; \quad \rho(\xi = -\infty) = \rho(\xi = +\infty) = 0. \qquad (14.30\text{a}-\text{c})$$

If the total energy of the system takes only certain values $E = E_n$, then the solution writes

$$\rho_n(\xi) = \sqrt{\frac{\Omega}{2\pi D}}\frac{1}{2^n n!}H_n^2(\xi)e^{-\xi^2} \qquad (14.31)$$

and

$$E_n = 2m_0 D\Omega\left(n + \frac{1}{2}\right), \qquad (14.32)$$

where $H_n(\xi)$ are the Hermite polynomials.

In particular, for $D_F = 2$ and $D = \hbar/2m_0$, we refined the results of quantum mechanics ([38], [39])

$$\Psi_n(x) = \left(\frac{m_o\Omega}{\pi\hbar}\right)^{1/4}\frac{1}{\sqrt{2^n n!}}e^{-\frac{m_0\Omega}{2\hbar}x^2}H_n\left(x\sqrt{\frac{m_0\Omega}{\hbar}}\right);$$

$$E_n = \left(n + \frac{1}{2}\right)\hbar\Omega, \quad n = 0, 1, 2, \ldots$$

where Ψ_n are the normalized wave functions of the stationary states and E_n are the discrete eigenvalues associated with those states.

According to our model, equations (14.27) and (14.31) have the following significance:

i) At the differentiable scale, the real (differentiable) part of the complex velocity field of the fractal fluid, $\mathbf{v}$, is null — equation (14.27). In agreement with equation (14.8a), $S = const.$, *i.e.* the fractal fluid is coherent (the fractal fluid particles have the same phase);

ii) At the non-differentiable scale, the imaginary (non-differentiable or fractal) part of the complex velocity field of the fractal fluid $\mathbf{u}$ is not zero; its expression is

$$u_x = D\frac{d(\ln\rho)}{dx} = D\frac{d(\ln\rho)}{d\xi}\frac{d\xi}{dx} = \sqrt{2D\Omega}\left[\frac{H_n'(\xi)}{H_n(\xi)} - \xi\right]$$

$$= \sqrt{2D\Omega}\left[2n\frac{H_{n-1}(\xi)}{H_n(\xi)} - \xi\right] \tag{14.33}$$

where we used the relationship [42]

$$H_n'(\xi) = \frac{dH_n(\xi)}{d\xi} = 2nH_{n-1}(\xi). \tag{14.34}$$

Then, the fractal one-dimensional potential

$$Q_n = -\frac{m_0 u_x^2}{2} - m_0 D\frac{du_x}{dx},$$

with

$$-\frac{m_0 u_x^2}{2} = -m_0 D\Omega\left[4n^2\frac{H_{n-1}^2(\xi)}{H_n^2(\xi)} + \xi^2 - 4n\xi\frac{H_{n-1}(\xi)}{H_n(\xi)}\right],$$

and

$$-m_0 D\frac{du_x}{dx} = -m_0 D\Omega\left[2n\frac{H_{n-1}'(\xi)H_n(\xi) - H_n'(\xi)H_{n-1}(\xi)}{H_n^2(\xi)} - 1\right]$$

$$= m_0 D\Omega\left[4n(n-1)\frac{H_{n-2}(\xi)}{H_n(\xi)} - 4n^2\frac{H_{n-1}^2(\xi)}{H_n^2(\xi)} - 1\right]$$

involves the explicit form

$$Q_n = -m_0 D\Omega \left[4n^2 \frac{H_{n-1}^2(\xi)}{H_n^2(\xi)} + \xi^2 - 4n\xi \frac{H_{n-1}(\xi)}{H_n(\xi)} \right.$$

$$\left. + 4n(n-1)\frac{H_{n-2}(\xi)}{H_n(\xi)} - 4n^2 \frac{H_{n-1}^2(\xi)}{H_n^2(\xi)} - 1 \right]$$

$$= -m_0 D\Omega \left[\xi^2 - 1 + 2n \frac{-2\xi H_{n-1}(\xi) + 2(n-1)H_{n-2}(\xi)}{H_n(\xi)} \right]$$

$$= m_0 D\Omega (2n + 1 - \xi^2), \tag{14.35}$$

where the following relation has been used [42]

$$H_n(\xi) - 2\xi H_{n-1}(\xi) + 2(n-1)H_{n-2}(\xi) \equiv 0. \tag{14.36}$$

One finally obtains

$$E_n = U + Q_n = m_0 D\Omega \xi^2 + m_0 D\Omega (2n + 1 - \xi^2)$$

$$\equiv 2m_0 D\Omega \left(n + \frac{1}{2} \right). \tag{14.37}$$

Therefore, in our model the harmonic oscillator in a fractal space-time is equivalent to a coherent fractal fluid whose particles move on stationary trajectories which satisfy the condition (14.37).

The identity of relations (14.32) and (14.37) show that the "observable" in the form of energy (generated by both the fractal component of the complex velocity field through the fractal potential and the external potential) is quantized. The zero value of the real (differentiable) part of the complex velocity field corresponds to a superconductor or a super-fluid type behavior of the fractal fluid, and the non-zero value of the imaginary (fractal) part selects, through condition (14.37), the stationary "trajectories" of the fluid particles. The momentum transfer is achieved only through the fractal component of the complex velocity field.

Written in the form $\Omega_n = (n + 1/2)\Omega$, condition (14.37) can simulate, in the case of a double layer self-structured plasma, a modulation of the layer oscillations under the influence of a given type of external signal (for details, see the voltage-current characteristic and the related comments in Ref. [43]).

14.5. Free particle in the fractal space-time

In the one-dimensional case, the fractal hydrodynamic equations (14.10a,b) and (14.11) take the form

$$m_0\left(\frac{\partial v}{\partial t} + v\frac{\partial v}{\partial x}\right) = -\frac{\partial}{\partial x}\left[-2m_0 D^2 \frac{1}{\sqrt{\rho}}\frac{\partial^2 \sqrt{\rho}}{\partial x^2}\right];$$

$$\text{(14.38a,b)}$$

$$\frac{\partial \rho}{\partial t} + \frac{\partial}{\partial x}(\rho v) = 0.$$

The initial state of the particle in a fractal space-time is specified by the velocity

$$v(x, t = 0) = c \tag{14.39}$$

and a Gaussian position distribution (distribution parameter α)

$$\rho(x, t = 0) = \frac{1}{\sqrt{\pi}\alpha}e^{-\left(\frac{x}{\alpha}\right)^2} = \rho_0(x). \tag{14.40}$$

This means that at $t = 0$, the center of the distribution $\rho(x)$ is at $\langle x \rangle_0 = 0$ and has velocity $\langle v \rangle_0 = c$. The boundary conditions are

$$v(x = ct, t) = c; \quad \rho(x = +\infty, t) = \rho(x = -\infty, t) = 0. \tag{14.41a, b}$$

At any $t > 0$ or $t < 0$, we have $\langle \partial Q/\partial x \rangle = 0$ (for details see [25]). Since $\langle x \rangle_0 = ct$, this suggests that, in the absence of external forces, equation (14.38a) can be separated into

$$\frac{\partial}{\partial x}\left(\frac{1}{\sqrt{\rho}}\frac{\partial^2 \sqrt{\rho}}{\partial x^2}\right) = \frac{2}{|\mu(t)|^2}(x - ct) \tag{14.42}$$

and

$$m_0\left(\frac{\partial v}{\partial t} + v\frac{\partial v}{\partial x}\right) = \frac{4m_0 D^2}{|\mu(t)|^2}(x - ct), \tag{14.43}$$

where $\mu(t)$ is a distribution parameter. Integration of equation (14.42) gives, under consideration of the boundary conditions

(14.41b), a quadratic solution in $(x - ct)$, *i.e.*

$$\rho(x,t) = \frac{1}{\sqrt{\pi\mu(t)}} \exp\left[-\frac{(x-ct)^2}{\mu(t)}\right]. \tag{14.44}$$

This function indeed satisfies the initial condition (14.40), if the initial value of $\mu(t)$ is chosen as

$$\mu(t = 0) = \alpha^2. \tag{14.45}$$

Introducing equation (14.44) into the continuity equation (14.38b) it is seen that, for $x = ct$, this equation writes

$$\frac{1}{2\mu}\frac{d\mu}{dt} = \left(\frac{\partial v}{\partial x}\right)_{x=ct}. \tag{14.46}$$

Then, the differential equation for $\mu(t)$ is obtained by applying the operation $(\partial/\partial x)_{x=ct}$ to equation (14.43)

$$\mu\frac{d^2\mu}{dt^2} - \frac{1}{2}\left(\frac{d\mu}{dt}\right)^2 = 8D^2. \tag{14.47}$$

The solution of equation (14.47) with the initial condition (14.45) (satisfying the requirement for $\rho(s,t)$ to be real; this condition is fulfilled if $\mu(t) = \mu(-t)$ — equation (14.44)) is

$$\mu(t) = \alpha^2 + \left(\frac{2D}{\alpha}t\right)^2. \tag{14.48}$$

According to equations (14.44) and (14.48), the density is Gaussian, with a time-dependent distribution parameter $\mu(t)$, and spreads with the "classical" particle velocity c,

$$\rho(x,t) = \frac{1}{\sqrt{\pi\left[\alpha^2 + \left(\frac{2D}{\alpha}t\right)^2\right]}} \exp\left[-\frac{(x-ct)^2}{\alpha^2 + \left(\frac{2Dt}{\alpha}\right)^2}\right]. \tag{14.49}$$

Similarly, integration of equation (14.43) with the initial condition (14.39), and the boundary condition (14.41a), gives the velocity field of the particle at the differentiable scale

$$v = \frac{c\alpha^2 + \left(\frac{2D}{\alpha}\right)^2 tx}{\alpha^2 + \left(\frac{2D}{\alpha}t\right)^2}.$$

(14.50)

Equations (14.49) and (14.50) represent the fractal-hydrodynamic solution of the free motion of a particle at the differentiable scale. At the non-differentiable scale, the velocity field of the particle

$$u(x,t) = D\frac{\partial \ln \rho}{\partial x} = -2D\frac{x - ct}{\alpha^2 + \left(\frac{2D}{\alpha}t\right)^2}$$

(14.51)

gets 'felt' through the fractal potential

$$Q = -\frac{m_0 u^2}{2} - m_0 D\frac{\partial u}{\partial x}$$

$$= -2m_0 D^2\frac{(x - ct)^2}{\left[\alpha^2 + \left(\frac{2D}{\alpha}t\right)^2\right]^2} + 2m_0 D^2\frac{1}{\alpha^2 + \left(\frac{2D}{\alpha}t\right)^2}$$

(14.52)

and the fractal force

$$F = -\frac{\partial Q}{\partial x} = 4m_0 D^2\frac{(x - ct)}{\left[\alpha^2 + \left(\frac{2D}{\alpha}t\right)^2\right]^2}.$$

(14.53)

Using the dimensionless coordinates

$$\tau = \omega t; \quad \xi = \frac{x}{\lambda}. \quad \gamma \equiv \left(\frac{\alpha}{\lambda}\right)^2; \quad \beta = \left(\frac{2D}{\alpha\omega\lambda}\right)^2,$$

(14.54a–d)

equations (14.49)-(14.53) give, successively:

- the normalized fractal probability density — see Figures 14.1a–c

$$\bar{\rho}(\xi,\tau) = \frac{\rho(\xi,\tau)}{\rho_0} = \frac{1}{\sqrt{\delta^2 + \beta^2\tau^2}}\exp\left[-\frac{(\xi - \tau)^2}{\delta^2 + \beta^2\tau^2}\right],$$

$$\rho_0 = \frac{1}{\sqrt{\pi\lambda}};$$

(14.55a,b)

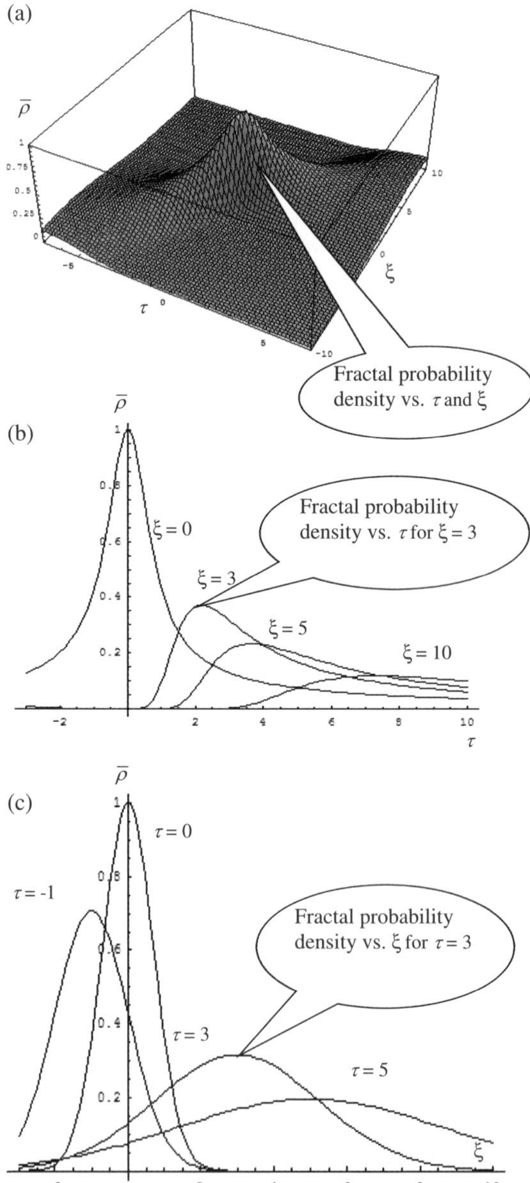

Figure 14.1. (a) The normalized fractal probability density $\bar{\rho}$ versus the fractal normalized position ξ and normalized time τ for $\delta = \beta = 1$; (b) The normalized fractal probability density $\bar{\rho}$ versus normalized time at different normalized fractal positions for $\delta = \beta = 1$; (c) The normalized fractal probability density $\bar{\rho}$ versus fractal normalized position at different normalized times for $\delta = \beta = 1$.

- the normalized differential velocity field — see Figures 14.2a–c

$$\bar{v}(\xi,\tau) = \frac{v(\xi,\tau)}{c} = \frac{\delta^2 + \beta^2\tau\xi}{\delta^2 + \beta^2\tau^2};\qquad (14.56)$$

- the normalized fractal velocity field — see Figures 14.3a–c

$$\bar{u}(\xi,\tau) = \frac{u(\xi,\tau)}{u_0} = -\frac{\xi-\tau}{\delta^2 + \beta^2\tau^2}\;,\quad u_0 = \frac{2D}{\lambda};\qquad (14.57a,b)$$

- the normalized fractal potential — see Figures 14.4a–c

$$\overline{Q}(\xi,\tau) = \frac{Q(\xi,\tau)}{Q_0} = -\frac{(\xi-\tau)^2}{(\delta^2 + \beta^2\tau^2)^2} + \frac{1}{\delta^2 + \beta^2\tau^2},\quad Q_0 = \frac{2m_0 D^2}{\lambda^2};$$
$$(14.58a,b)$$

- and the normalized fractal force — see Figures 14.5a–c

$$\overline{F}(\xi,\tau) = \frac{F(\xi,\tau)}{F_0} = \frac{\xi-\tau}{(\delta^2 + \beta^2\tau^2)^2},\quad F_0 = \frac{4m_0 D^2}{\lambda^3}.\qquad (14.59a,b)$$

Interpreting these figures, one can observe that:

i) there is an alteration of the Gaussian profile of the normalized fractal probability density both with τ (for different values of ξ), and with ζ (for different values of τ);

ii) if the profile symmetry of the normalized differential velocity field maintains only at the origin of the fractal space-time normalized coordinates, it disappears both with τ (for different values of ξ) and with ξ (for different values of τ);

iii) if, initially, the profile symmetry of the normalized fractal velocity field is broken at the origin of the fractal space-time normalized coordinates, it shall be reconstructed only with τ (for different values of ξ);

iv) the profile symmetry of the normalized fractal potential is broken at the origin of the fractal space-time normalized coordinates and it will be reconstructed only with τ (for different values of ξ);

v) the profile symmetry of the normalized fractal force is broken at the origin of the fractal space-time normalized coordinates and it will be reconstructed only with τ (for different values of ξ).

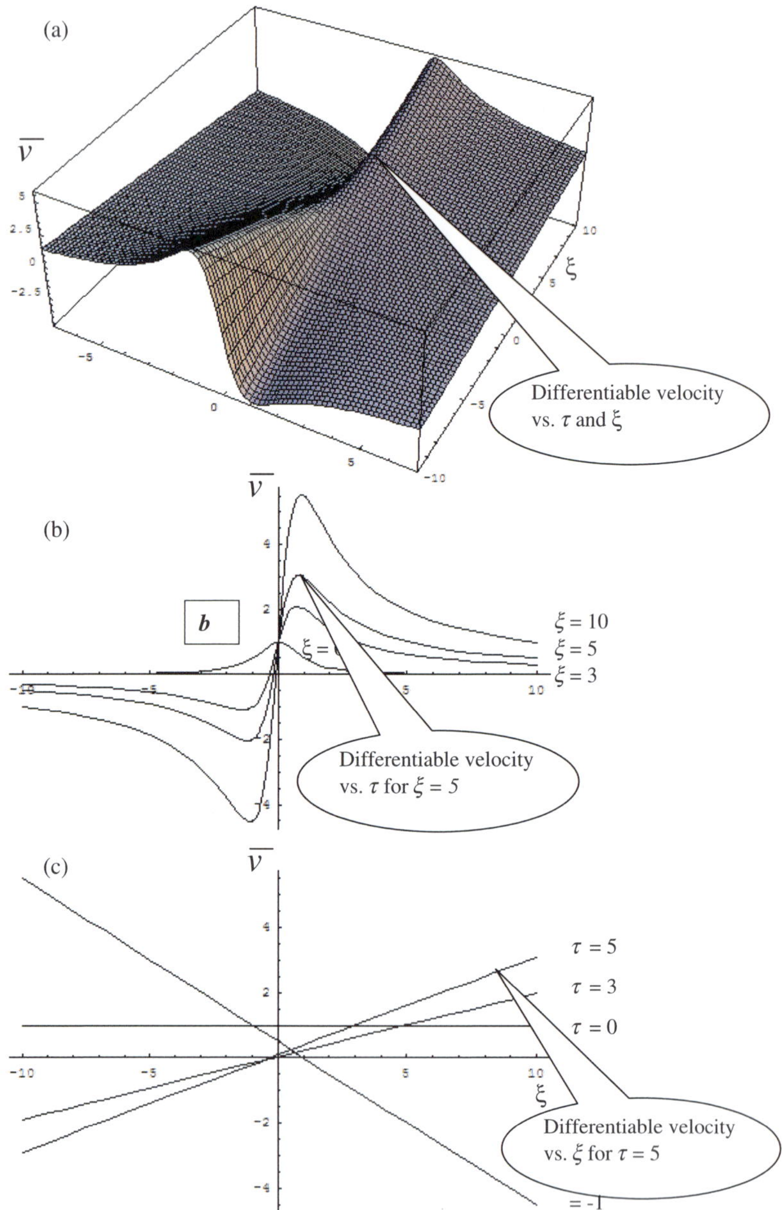

Figure 14.2.   (a) The normalized differentiable velocity field $\bar{v}$ versus the fractal normalized position ξ and normalized time τ for $\delta = \beta = 1$; (b) The normalized differentiable velocity field $\bar{v}$ versus normalized time at different normalized fractal positions for $\delta = \beta = 1$; (c) The normalized differentiable velocity field $\bar{v}$ versus the fractal normalized position at different normalized times for $\delta = \beta = 1$.

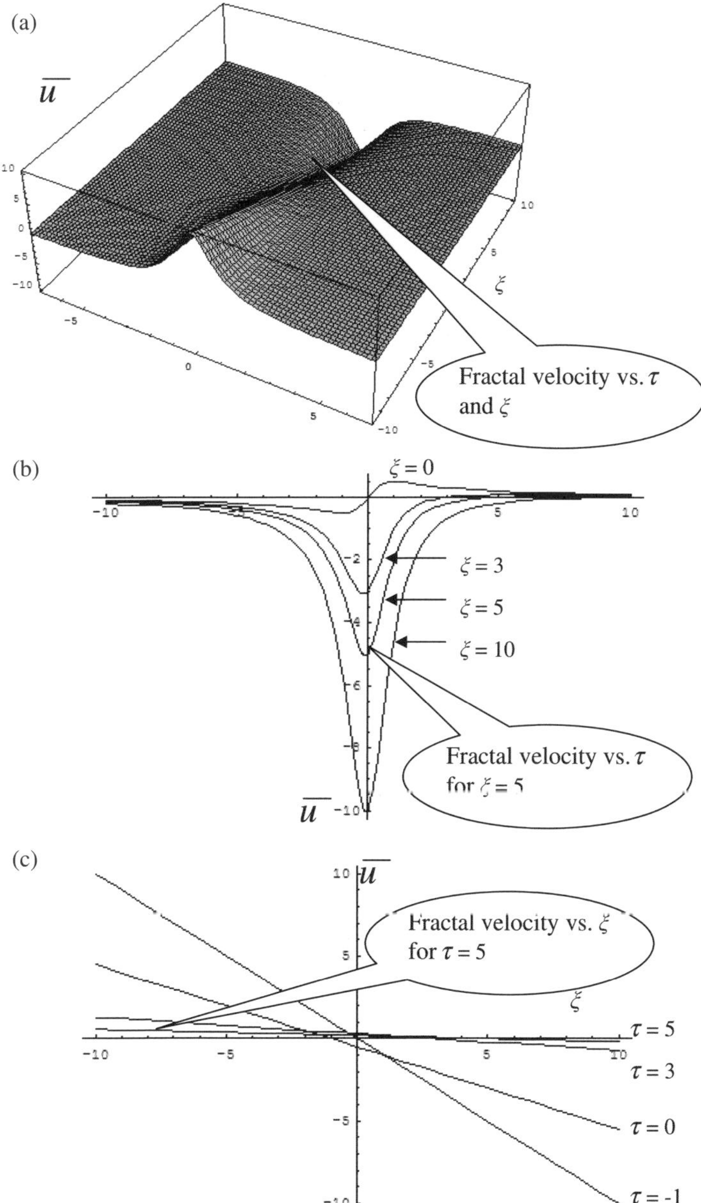

Figure 14.3. (a) The normalized fractal velocity field $\bar{u}$ versus the fractal normalized position ξ and normalized time τ for $\delta = \beta = 1$; (b) The normalized fractal velocity field $\bar{u}$ versus normalized time at different normalized fractal positions for $\delta = \beta = 1$; (c) The normalized fractal velocity field $\bar{u}$ versus the fractal normalized position at different normalized times for $\delta = \beta = 1$.

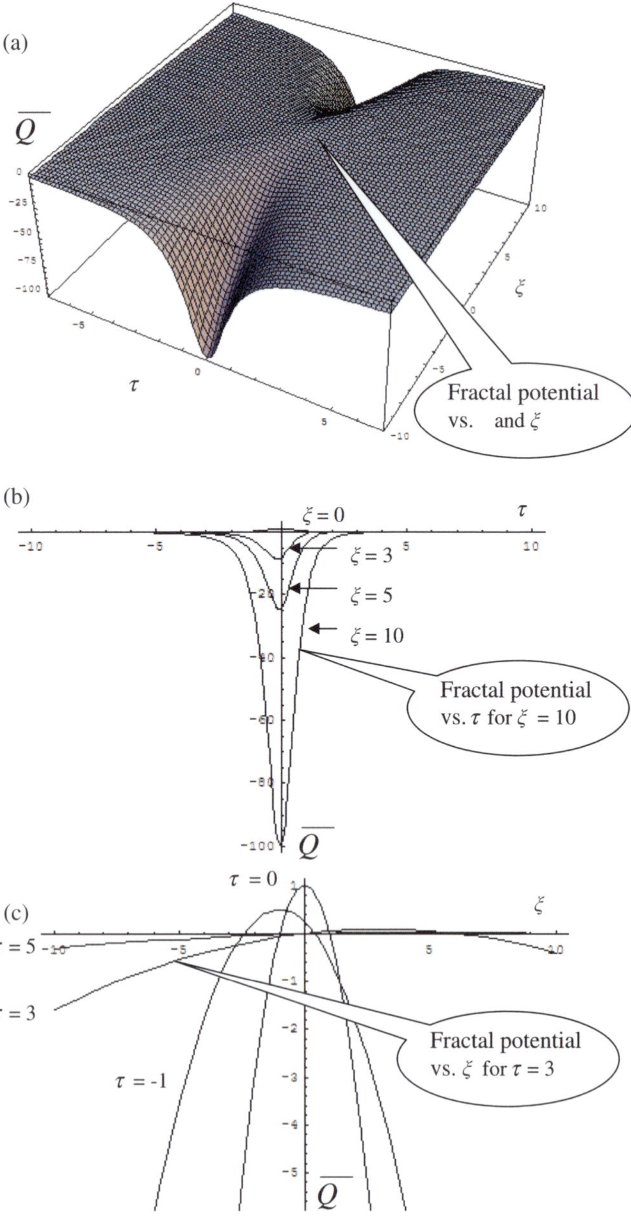

Figure 14.4.   (a) The normalized fractal potential $\overline{Q}$ versus the fractal normalized position ξ and normalized time τ for $\delta = \beta = 1$; (b) The normalized fractal potential $\overline{Q}$ versus normalized time at different normalized fractal positions for $\delta = \beta = 1$; (c) The normalized fractal potential $\overline{Q}$ versus the fractal normalized position at different normalized times for $\delta = \beta = 1$.

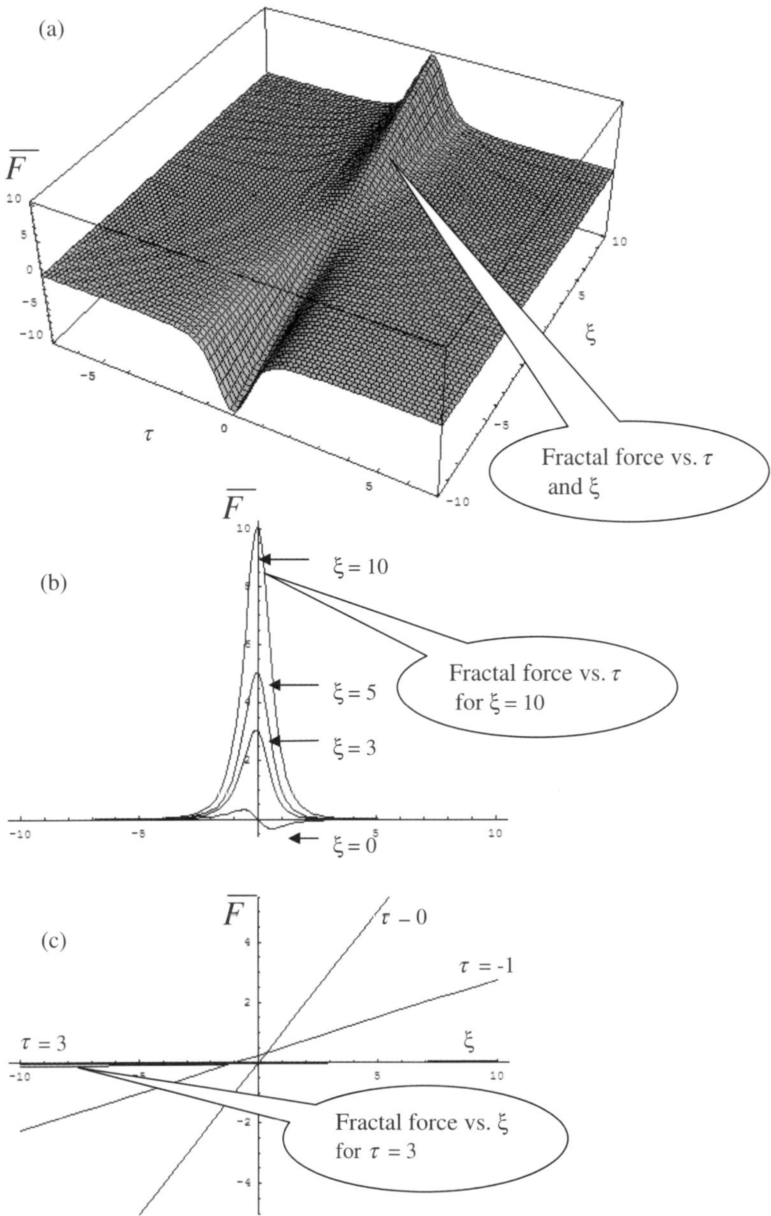

Figure 14.5. (a) The normalized fractal force $\overline{F}$ versus the fractal normalized position ξ and normalized time τ for $\delta = \beta = 1$; (b) The normalized fractal force $\overline{F}$ versus normalized time at different normalized fractal positions for $\delta = \beta = 1$; (c) The normalized fractal force $\overline{F}$ versus the fractal normalized position at different normalized times for $\delta = \beta = 1$.

The observable motion of the particle is uniform, $\langle v \rangle = c$, while the associated motions at both the differentiable scale — see equation (14.50), and non-differentiable one — see equation (14.51), are inhomogeneous in x and t, the fractal space being provided with a fractal force (14.53).

The fractal forces on the semi-spaces $-\infty \leq x \leq \langle x \rangle$ and $\langle x \rangle \leq x \leq +\infty$. with $\langle x \rangle = const.$, compensate each other

$$\left\langle m_0 \frac{dv}{dt} \right\rangle_{x=-\infty}^{\langle x \rangle} = \left\langle m_0 \frac{dv}{dt} \right\rangle_{\langle x \rangle}^{x=+\infty}.$$

This means that the particle in free motion in the fractal space-time polarizes the "vacuum" behind itself, $x \leq ct$, and ahead of itself, $x \geq ct$, in such a way that the resulting fractal forces are distributed symmetrically with respect to the plane through the observable particle position $\langle x \rangle = ct$ at any time t — see the symmetry of the curves given in Figures 14.1–14.5. Our results generalize those obtained in [39]. Therefore, in our model the free time-dependent particle in a fractal space-time is equivalent to an incoherent fractal fluid. Both the differentiable and the fractal components of the complex velocity field are inhomogeneous in fractal coordinates due to the action of a fractal force. A momentum transfer on both velocity components exists, and the "observable" movement, *i.e.* the uniform movement, is naturally given by means of a specific mechanism of vacuum polarization (the fractal component of the complex velocity field and the fractal force become simultaneously zero). Discarding the symmetry and considering only the positive part of the curves from Figures 14.1–14.5 allow us to build a theoretical model which can describe the expansion of laser produced plasma [44]. We note that similar results concerning the space-time dependencies were obtained by Mora [45], Murakami [46] etc.

14.6. Fractal conservation laws

Let us apply the complex operator $\hat{\partial}/\partial t$ — see equation (14.1n) — to an arbitrary fractal function $\varepsilon = \varepsilon(r, t)$ (see [5], [10]. [11] for details).

The result is

$$\frac{\hat{\partial}\varepsilon}{\partial t} = \frac{\partial \varepsilon}{\partial t} + \mathbf{V} \cdot \nabla \varepsilon - i\frac{\lambda^2}{2\tau}\left(\frac{dt}{\tau}\right)^{\left(\frac{2}{D_F}\right)-1} \Delta \varepsilon = 0, \tag{14.60}$$

and, by separating the real and imaginary parts,

$$\frac{\partial \varepsilon}{\partial t} + \mathbf{v} \cdot \nabla \varepsilon = 0; \quad -\mathbf{u} \cdot \nabla \varepsilon = \frac{\lambda^2}{\tau}\left(\frac{dt}{\tau}\right)^{\left(\frac{2}{D_F}\right)-1} \Delta \varepsilon. \tag{14.61a, b}$$

Consequently, at the differentiable scale, the local temporal variation $\partial \varepsilon/\partial t$ and the term $\mathbf{v} \cdot \nabla \varepsilon$ are equal, while on the non-differentiable scale, the terms $\mathbf{u} \cdot \nabla \varepsilon$ and $\Delta \varepsilon$ compensate each other. In particular, for $\mathbf{v} = \mathbf{u}$ ("synchronic" movements at different scales), equations (14.61a,b) yield the diffusion-type equation

$$\frac{\partial \varepsilon}{\partial t} = \frac{\lambda^2}{\tau}\left(\frac{dt}{\tau}\right)^{\left(\frac{2}{D_F}\right)-1} \Delta \varepsilon. \tag{14.62}$$

Such an equation is implied by the Fourier type law

$$\mathbf{j}(\varepsilon) = \frac{\lambda^2}{\tau}\left(\frac{dt}{\tau}\right)^{\left(\frac{2}{D_F}\right)-1} \nabla \varepsilon, \tag{14.63}$$

where $\mathbf{j}(\varepsilon)$ is the current density. Therefore, equations (14.62) and (14.63) describe a fractal mechanism of conduction type at different scales.

Moreover, multiplying equation (13.10b) by ε, *i.e.*

$$\frac{\partial(\rho\varepsilon)}{\partial t} + \nabla \cdot (\rho\varepsilon\mathbf{v}) = \rho\left(\frac{\partial \varepsilon}{\partial t} + \mathbf{v} \cdot \nabla \varepsilon\right) \tag{14.64}$$

and taking into account equation (14.61a), the conservation law for ε is found in the form

$$\frac{\partial(\rho\varepsilon)}{\partial t} + \nabla \cdot (\rho\varepsilon\mathbf{v}) = 0. \tag{14.65}$$

In particular, if ε is the energy density of a fluid [47], $\varepsilon = e + (p/\rho) + v^2/2$, then the "classical" form of the energy conservation law is obtained.

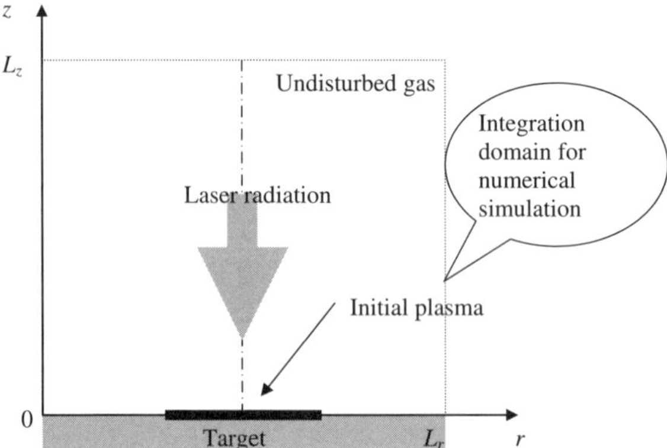

Figure 14.6.   Integration domain used for the numerical simulation of the laser-produced aluminum plasma expansion.

Let us now apply the previous considerations in the numerical simulations of plasma expansion produced by the laser ablation. The plasma expansion is supposed to have axial symmetry and a system of cylindrical coordinate induced in the region above the target surface is used (Figure 14.6). The z-axis coincides with the laser beam axis, being directed along the outer normal to the target surface. The plasma evolution is described with the following assumptions:

i) the plasma is in the state of local thermo-dynamical equilibrium and satisfies the quasi-neutrality condition;

ii) the expansion is described in the approximation of a non-viscous, non-thermo-conducting gas;

iii) the release of energy by thermal radiation is neglected, and the ideal gas equations of state are considered;

iv) the source term is introduced through the boundary conditions.

Under such circumstances, the two-dimensional gas dynamics is described by the Equations (14.10a,b) with $\nabla Q = -\nabla p/\rho$ and the energy balance equation (14.65), that is, explicitly

$$\frac{\partial n}{\partial t} + \frac{1}{r}\frac{\partial}{\partial r}(rnu) + \frac{\partial}{\partial z}(nv) = 0;$$

$$\frac{\partial (nu)}{\partial t} + \frac{1}{r}\frac{\partial}{\partial r}(rnu^2) + \frac{\partial}{\partial z}(nvu) = -\frac{\partial p}{\partial r};$$

$$\frac{\partial(nv)}{\partial t} + \frac{1}{r}\frac{\partial}{\partial r}(rnuv) + \frac{\partial}{\partial z}(nv^2) = -\frac{\partial p}{\partial z};$$

$$\frac{\partial(ne)}{\partial t} + \frac{1}{r}\frac{\partial}{\partial r}(rnue) + \frac{\partial}{\partial z}(nve) = -p\left[\frac{1}{r}\frac{\partial}{\partial r}(ru) + \frac{\partial v}{\partial z}\right].$$

$$(14.66\text{a-d})$$

Here t is the time, r and z the spatial coordinates, n the atoms density, while u and v are the velocity vector components.

For the numerical integration, the following initial and boundary conditions are considered:

i) The box integration domain is initially filled with undisturbed gas,

$$t = 0, \quad u = v = 0, \quad n = n_0,$$
$$T = T_0, \quad 0 \leq (r \times z) \leq (L_r \times L_z),$$
$$(14.67\text{a–e})$$

where T is the temperature.

ii) The interaction of the laser beam with the target produces a plasma source located on the target surface, which is assumed to have a Gaussian space-time profile

$$z = 0 \; : \; u = v = 0, \quad T = T_{plasma},$$
$$(14.68\text{a–d})$$

$$n = n_{max}\exp\left[-\frac{(t-\tau)^2}{(\tau_L/2)^2}\right]\exp\left[-\frac{(r-L_r/2)^2}{(d_L/2)^2}\right],$$

with d_L, τ_L similarly to the laser beam space-time full widths, and $T_{plasma} = 11.8\,eV$ the initial plasma temperature. We mention that the ablation takes place only into a region with characteristic diameter of about $100\,\mu m$. The maximum atoms density n_{max} is taken according to the critical electron density ($n_{ec} = 3.9 \cdot 10^{21}\,cm^{-3}$ [48]) at the laser wavelength ($\lambda = 532\,nm$), and the average ions charge state $\overline{Z} = 2$ (for details see [49]).

iii) The symmetry condition,

$$r = 0, L_r \ : \ u(0) = u(L_r), \quad \nu(0) = \nu(l_r),$$

$$n(0) = n(L_r), \quad T(0) = T(L_r),$$

$$(14.69\text{a-e})$$

and the undisturbed gas is considered on the upper boundary,

$$z = L_z \ : \ u = v = 0, \quad n = n_0, \quad T = T_0. \qquad (14.70\text{a-d})$$

The system of equations (14.66a–c), with conditions (14.67)–(14.70), are numerically solved using finite differences method [50], with tho following values of the parameters: $L_r = L_z = 300\,\mu m$, $\tau_L = 10\,ns$, $d_L = 100\,\mu m$, $n_{max} = 1.95 \cdot 10^{21}\,cm^{-3}$, $n_0 = n_{max}/1000$, $T_0 = 0.1\,eV$.

Figures 14.7(a–c) displays the 2D-contour curves of the total atom density at the time moments: (a) $t = 5\,ns$, (b) $t = 7\,ns$, $(c)t = 9\,ns$, as being obtained by means of numerical simulation.

These results show that he shape of the plasma core is in agreement with the experimental observations [51]. Moreover, the "mushroom" type structure given in Figures 14.7(a–c) and the vertical motion arising at the plume periphery are also in a good agreement with the experimental observations ([52], [53]). Consequently, the numerical simulations can describe quantitatively well the behavior of plasma at the early stages of its evolution (at nanosecond time scale).

Similar numerical results. but at a different time scale, are also given in [51], where the Gaussian space-time profile of the laser beam is considered. Nevertheless, we mention that the critically self-organized plasma processes require non-Gaussian profiles. For these situations some numerical simulations, together with correspondences between "approximations" induced through both types of profiles, are specified in [54].

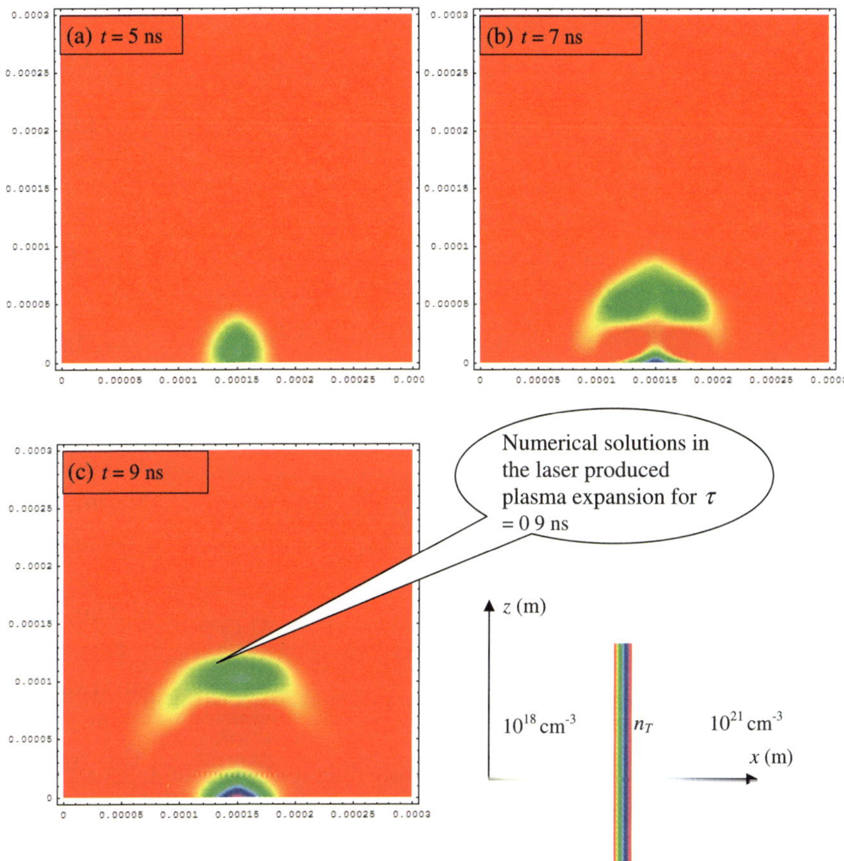

Figure 14.7. The numerical solutions for the total atom density (n_T) in the laser produced plasma expansion at the moments of time (a) $t = 5\,ns$; (b) $t = 7\,ns$; (c) $t = 9\,ns$.

14.7. Conclusions

In the light of the above investigation, we can draw the following concluding remarks:

i) Considering that the motion of the particles takes place on continuous but non-differentiable curves, *i.e.* on fractals, an extended fractal hydrodynamic model in a constant arbitrary fractal dimension D_F, as well as the second order terms in the equation of motion for the complex velocity field, are established;

ii) In the case of such movements, a Navier-Stokes-type equation with an imaginary viscosity coefficient and, from here, in the particular case of irrotational movement, a Schrödinger-type equation are obtained. The standard Schrödinger equation is found as a particular case of irrotational movement at Compton scale, in the constant fractal dimension $D_F = 2$;

iii) The extended fractal hydrodynamic model follows from a Navier-Stokes-type equation by separating the real and the imaginary parts of the complex velocity field. The model is described through a momentum transport equation and a conservation law of the probability density;

iv) Using our model in its hydrodynamic version, some applications are given. Also, for the static systems of potential box-type or linear harmonic oscillator-type, the complex velocity field of the fractal fluid proves to be essential: the zero value of the real (differentiable) part specifies the fractal fluid coherence (superconductor or super-fluid type behavior), while the non-zero value of the imaginary (non-differentiable or fractal) part selects, through some quantification condition, the stationary trajectory of the fractal fluid particles. Moreover, the momentum transfer is achieved only through the fractal component of the complex velocity field of the fractal fluid, and the same fractal component imposes totally (see particle in a box in a fractal space-time) or partially (see the harmonic oscillator) the "observable" in the form of energy which is quantized;

v) It also emerges a new physical meaning associated with the concept of particle: various properties of "particles" can be reduced to the geometric structures of the (non-differentiable) geodesics of the fractal space-time;

vi) For the free time-dependent particle in a fractal space-time, both the differentiable and the fractal components of the complex velocity field are inhomogeneous in fractal coordinates, due to the action of a fractal force. There exists a momentum transfer on both velocity components and the "observable" movement, *i.e.* the uniform movement is naturally given by means of a specific mechanism of vacuum polarization. More precisely, the particle in free motion in the fractal space-time polarizes the vacuum behind itself and ahead of itself. The resulting fractal force field is distributed symmetrically with respect to the plane through the observable particle position. On this plane the fractal velocity field and the fractal force are null at any time. Discarding the symmetry and considering only the positive part of the curves shown in Figures 14.1–14.5 allow us to build a theoretical model which can describe the expansion of the laser produced plasma;

vii) The conservation laws, and particularly the energy conservation law, are deduced for the fractal quantities. In such a context, using the fractal hydrodynamic equations and the energy conservation law, the numerical simulations of plasma expansion produced by laser ablations are obtained. These numerical simulations can describe quantitatively well the plasma behavior at the early stages (at nanoseconds time scale);

viii) Since the transport phenomena in the "matter" imply different space time scales, according to the previous results, some new transport mechanisms are obtained. Thus, for a coherent fractal fluid, one obtains the momentum transport through the fractal velocity field. In case of an incoherent fractal fluid, a fractal mechanism of conduction type for synchronal movement at different scales, and a convection type mechanism in the absence of the "synchronal" movements, are also found.

We note that it is possible to obtain the same results using a fractal action or fractional calculus of variations ([55]–[59]).

References (Chapter 14)

[1] B. Madelbrot, *The Fractal Geometry of Nature* (Freeman, San Francisco, 1982)

[2] E. Nelson, *Quantum Fluctuations* (Princeton University Press, Princeton, NY, 1985)

[3] J. Feder, A. Aharony, *Fractals in Physics* (North Holland, Amsterdam, 1990)

[4] J. F. Gouyet, *Physique et Structures Fractals* (Masson, Paris, 1992)

[5] L. Nottale, *Fractal Space-Time and Microphysics: Towards a Theory of Scale Relativity* (World Scientific, Singapore, 1993)

[6] L. Nottale, *Scale Realtivity and Fractal Space-Time. A New Approach to Unifying Relativity and Quantum Mechanics*, (Imperial Colege Press, 2011)

[7] J. Argyris, C. Ciubotariu, G. Mattatis, *Chaos, Solitons and Fractals* **12**, 1 (2001)

[8] P. Weibel, G. Ord, G. Rssler (Editors) *Space-time Physics and Fractality* (Vienna, New York, Springer 2005)

[9] C. P. Cristescu, *Nonlinear Dynamics and Chaos in Science and Engineering* (Bucharest, Academy Publishing House, 2008)

[10] J. Cresson, F. Ben Adda, *Chaos, Solitons and Fractals* **19**, 1323 (2004)

[11] J. Cresson, *J. Math. Anal. Appl.* **307**, 48 (2005)

[12] G. N. Ord, *J. Phys. A* **16**, 1869 (1983)

[13] N. Laskin, *Fractional Quantum Mechanics*, Phys. Rev. E 2000; 62:3135

[14] N. Laskin, *Fravtional Quantum Mechanics and Levy Path Integrals*, Phys. Lett. A 2000; 268:298

[15] L. Nottale, *Astron. Astrophys.* **327**, 867 (1997)

[16] L. Nottale, *Chaos Solit. Fract.* **9**, 1051 (1980)

[17] L. Nottale, *Chaos Solit. Fract.* **10**, 459 (1999)

[18] L. Nottale, *Chaos Solit. Fract.* **16**, 539 (2003)

[19] D. Da Rocha, L. Nottale, *Chaos Solit. Fract.* **16**, 565 (2003)

[20] L. Nottale, *Chaos Solit. Fract.* **25**, 797 (2005)

[21] L. Nottale, M. N. Célérier, T. Lehner, *J. Math. Phys.* **47**, 032303 (2006)

[22] M. N. Célérier, L. Nottale, *J. Phys. A: Math. Gen.* **37**, 931 (2004)

[23] I. Gottlieb, M. Agop, G. Ciobanu, A. Stroe, *Chaos Solit. Fract.* **30**, 380 (2006)

[24] M. Agop, P. D. Ioannou, P. Nica, *J. Math. Phys.* **46**, 062110 (2005)

[25] M. Agop, P. E. Nica, P. D. Ioannou, A. Antici, V. P. Paun, *Eur. Phys. J. D* **49**, 239248 (2008)

[26] M. Agop, P. Nica, M. Girtu, *Gen. Relat. Gravit.* **40**, 35 (2008)

[27] M. Agop, P. Nica, P. D. Ioannou, O. Malandraki, I. Gavanas-Pahomi, *Chaos Solit. Fract.* **34**, 1704 (2007)

[28] A. G. Agnese, R. Festa, *Phys. Lett. A* **227**, 165, 1997

[29] E. Schrödinger, *Collected Papers on Wave Mechanics* (W. M. Deans, London, 1928)

[30] R. P. Feynman, A. R. Hibbs, *Quantum Mechanics and Path Integrals* (Mc Graw Hill, New York, 1965)

[31] F. Halbwachs, *Theorie Relativiste des Fluid a Spin* (Gauthier-Villars, Paris, 1960)

[32] J. Argyris, C. Marin, C. Ciubotariu, *Physics of Gravitation and the Universe* (Tehnica-Info and Spiru Haret Publishing Houses, Iasi, 2006)

[33] E. A. Jackson, *Perspectives in Nonlinear Dynamics*, Vol. I, II (Cambridge University Press, Cambridge, 1991)

[34] A. V. Oppenheim, R. W. Schafer, *Discrete-Time Signal Processing* (Prentince-Hall, 1989)

[35] M. Agop, C. Murgulet, *Nonlinear Dynamics, Ball Lightning and Cosmic Structures* (Ars Longa Publishing House, Iasi, 2006)

[36] V. Chiroiu, P. Stiuca, L. Munteanu, S. Danescu, *Introduction in Nanomechanics* (Romanian Academy Publishing House, Bucharest, 2005)

[37] D. K. Ferry, S. M. Goodnick, *Transport in Nanostructures* (Cambridge University Press, Cambridge, 1997)

[38] L. E. Ballentine, *Quantum mechanics. A Modern Development* (World Scientific, Singapore, 1998); A. C. Phillips, *Introduction to Quantum Mechanics* (John Wiley and Sons, New York, 2003)

[39] H. E. Wilhem, Phys. Rev. D **1**, 2278 (1970)

[40] L. Conde, L. Leon, *Physics of Plasmas* **1**, 2441, 1994

[41] C. Ionita, D. G. Dimitriu, R. Schrittwieser, *International Journal of Mass Spectrometry* **233**, 343, 2004

[42] A. Nikitorov, V. Ouvarov, Éléments de la Théorie des Functions Spéciales (Mir, Moskow, 1974)

[43] I. Alcaide, P. C. Balam, L. Conde, C. Ionita, R. Schrittwieser, Contrib. Plasma Phys. 43 (5–6), 373, 2003

[44] P. Nica, P. Vizureanu, M. Agop, S. Gurlui, C. Focsa, N. Forna, P. D. Ioannou, Z. Borsos, *Jpn. J. Appl. Phys.* **49**, 2009

[45] P. Mora, *Phys. Rev. Lett.* **90**, 185002, 2003

[46] M. Murakami, Y. G. Kang, K. Nishihara, H. Nishimura, *Physics of Plasmas* **12**, 062706, 2005

[47] L. Landau, E. Lifshitz, *Fluid Mechanics* (Butterworth-Heinemann, Oxford, 1987)

[48] L. Spitzer, *Physics of Fully Ionized Gases* (Wiley, New York, 1962)

[49] I. C. E. Turcu, J. B. Dance, *X-Rays from Laser Plasmas* (Wiley, Chichester, U. K., 1998)

[50] O. C. Zienkievicz, R L. Taylor, *The Finite Element Method* (McGraw-Hill, New York, 1991)

[51] S. Gurlui, M. Agop, P. Nica, M. Ziskind, C. Focsa, *Phys. Rev. E* **78**, 062706, 2008

[52] S. S. Harilal, C. V. Bindhu, M. S. Tillack, F. Najmabadi, A. C. Gaeris, *J. Appl. Phys.* **93**, 2380, 2003

[53] A. V. Bulgakov, N. M. Bulgakova, *J. Phys. D* **31**, 693, 1998

[54] P. Cristescu, *Nonlinear Dynamics and Chaos. Theoretical Fundaments and Applications* (Academy Publishing House, Bucharest, 2008)

[55] A. R. El-Nabulsi, *Chaos, Solitons and Fractals* **42**, 2384, 2009

[56] A. R. El-Nabulsi, *Chaos, Solitons and Fractals* **42**, 2924, 2009

[57] B. M. Hambly, M. L. Lapidus, *Trans. Amer. Math. Soc.* **358**, 285, 2006

[58] A. Arneodo, F. Argoul, E. Bacry, J. F. Muzy, *Phys. Rev. Lett.* **68**, 3456, 1992

[59] A. Arneodo, F. Argoul, J. F. Muzy, M. Tabard, E. Bacry, *Fractals* **1**, 629, 1993

Chapter 15

Theory of Fractional Scale Relativity
and Some Applications

15.1. Introduction

In the investigation to follow we shall present some results applied
for Scale Relativity Theory in the light of El Nabulsi's research [1].
The idea of a fractal operator (derivative and integral) is not new;
it is as old as the definition of an integer order derivative. In general, there is a strong connection between Levi's random processes
and fractional calculus ([2]–[4]). Fractal operators are not limited
to classical or Hamiltonian dynamics. In 1951 Kac [5] discussed the
possibility of extending the Feynman path integrals over the paths
of Levy dynamics, and later Laskin [6] studied the generalization
of quantum mechanics, namely Schrödinger equation. The idea is
that the paths integrals over Levy trajectories lead to the fractional
Schrödinger equation in the same way as the classical one. The space
derivative operator is fractal of order $0 < \alpha \leq 2$ ([4]–[7]).

15.2. Basic mathematical tools and ideas

Fractal Euler-Lagrange equations

Consider a smooth n-dimensional manifold M and let $L : TM \times M \times \Re \to \Re$ be the smooth Lagrangian function. For any smooth path $q :
[a, b] \to M$ satisfying fixed boundary conditions $q(a) = q_0$ and $q(b) =
q_b$, the fractional functional associated with $(q, \dot{q}, \tau) \to L(q, \dot{q}, \tau)$ is

269

defined by ([8]–[19])

$$S^{(\alpha,\beta)}_{(\gamma)}[q](t) = \frac{1}{\Gamma(\alpha)} \int_0^t L(D^{(\alpha,\beta)}_{(\gamma)} q(\tau), q(\tau), \tau)(t-\tau)^{\alpha-1}\, d\tau;$$

$$[a,t] \subseteq [a,b] \subset \Re. \tag{15.1}$$

Here $D^{(\alpha,\beta)}_{(\gamma)}$ is the fractional derivative operator of order (α,β), $0 < \alpha, \beta < 1$, $a,b,t \in \Re$, $a < t$, defined by Cresson as

$$D^{(\alpha,\beta)}_{(\gamma)} = \frac{1}{2}\left[D^{(\alpha)}_{(a+)} - D^{(\beta)}_{(b-)} \right] + \frac{i\gamma}{2}\left[D^{(\alpha)}_{(a+)} + D^{(\beta)}_{(b-)} \right] \tag{15.2}$$

where $\gamma \in C$, $i = \sqrt{-1}$, and $\Gamma(\alpha) = \int_0^\infty t^{\alpha-1}\exp(-t)\,dt$ is the Euler gamma function. For any smooth function $f(t)$ we admit the following properties

$$D^{(\alpha)}_{(a+)}f(t) = \frac{1}{\Gamma(1-\alpha)}\frac{d}{dt}\int_a^t f(\tau)(t-\tau)^{-2}d\tau;$$

$$\tag{15.3a,b}$$

$$D^{(\beta)}_{(b-)}f(t) = \frac{1}{\Gamma(1-\beta)}\left(-\frac{d}{dt}\int_t^b f(\tau)(\tau-t)^{-2}\,d\tau \right),$$

which are the left and right Riemann-Liouville fractional derivatives of order (α,β); $0 < \alpha, \beta < 1$, augmented by the following properties

$$\int_a^b D^{(\alpha,\beta)}_{(\gamma)} f(t)g(t)dt = -\int_a^b f(t)D^{(\beta,\alpha)}_{-(\gamma)} g(t)dt, \quad a \le t \le b; \tag{15.4}$$

$$\int_a^b D^{(\alpha)}_{(a)} f(t)g(t)dt = (-1)^\alpha \int_a^b f(t)D^{(\alpha)}_{(b-)} g(t)dt; \tag{15.5}$$

$$D^{(\alpha)}_{(a)} D^{(\beta)}_{(a)} f(t) = D^{(\alpha+\beta)}_{(a)} f(t) - \sum_{i=1}^k D^{(\alpha-1)}_{(a)} f(t)\Big|_{t=a} \frac{(t-a)^{-\alpha-i}}{\Gamma(1-\alpha-i)},$$

$$0 \le k-1 \le q \le k, \tag{15.6}$$

provided that $f(a) = f(b) = 0$ and $g(a) = g(b) = 0$; k in Eq.(14.6) is a whole number. For $\gamma = i$, $D^{(\alpha,\beta)}_{(\gamma)} = -D^{(\beta)}_{(b-)}$, and for $\gamma = -i$, $D^{(\alpha,\beta)}_{(\gamma)} = D^{(\alpha)}_{(a+)}$, τ is the intrinsic time, and t the observer's time. For $\alpha \to 1$

and $\beta \to 1$, one obtains $D^{(\alpha,\beta)}_{(\gamma)} = d/dt$ and the action reduces to its standard and normal form. It can be proved that the corresponding fractional Euler-Lagrange equations then take the special form

$$\frac{\partial L(D^{(\alpha,\beta)}_{(\gamma)} q(\tau), q(\tau), \tau)}{\partial q} - D^{(\beta,\alpha)}_{(-\gamma)} \left(\frac{\partial L(D^{(\alpha,\beta)}_{(\gamma)} q(\tau), q(\tau), \tau)}{\partial \dot{q}} \right)$$

$$= \frac{1-\alpha}{1-\tau} \frac{\partial L(D^{(\alpha,\beta)}_{(\gamma)} q(\tau), q(\tau), \tau)}{\partial \dot{q}}. \quad \forall \tau = (a, t). \tag{15.7}$$

The previous arguments can be repeated, *mutatis mutandis*, to the N-dimensional problem where the admissible paths are smooth functions $q : \Omega \subset \Re^N \to M$ satisfying the given Dirichlet's boundary conditions on $\partial\Omega$. The Lagrangian function $(q_{x_1}, \ldots, q_{x_N}, q, x_1, \ldots, x_N) \to L(q_{x_1}, \ldots, q_{x_N}, q, x_1, \ldots, x_N)$ is supposed to be sufficiently smooth with respect to all its arguments. The N-dimensional action is defined by

$$S^{(\alpha,\beta)}_{(\gamma)}[q](\xi) = \frac{1}{\Pi_{i=1}^{N}\Gamma(\alpha_i)} \int \cdots \int_{\Omega(\xi)} L(\nabla^{(\alpha,\delta)}_{(\gamma i)} q(x), q(x), x)$$

$$\times \Pi_{i=1}^{N}(\xi_i - x_i)^{\alpha_i - 1} dx, \tag{15.8}$$

where $\nabla^{(\alpha,\delta)}_{(\gamma i)} = (D^{\alpha_1 \delta_1}_{\gamma x_1}, \ldots, D^{\alpha_n \delta_n}_{\gamma x_N})$, $\alpha = (\alpha_1, \ldots, \alpha_N)$, $\delta = (\delta_1, \ldots, \delta_N)$ $0 < \alpha_i < 1$, $(i = 1, \ldots, N)$, and $dx = dx_1, \ldots, dx_N$.

The N-dimensional associated fractional Euler-Lagrange equation is given by

$$\sum_{j=1}^{N} \left[D^{(\delta_j X_j)}_{(\gamma)j x_j} \left(\frac{\partial L}{\partial q_{x_j}} \right) + \frac{1-\alpha_j}{\xi_j - \alpha_j} \left(\frac{\partial L}{\partial q_{x_j}} \right) \right] - \frac{\partial L}{\partial q} = 0, \tag{15.9}$$

where all partial derivatives of the Lagrangian are evaluated at $(\nabla^{(\alpha,\beta)}_{(\gamma)} q(x), q(x), x)$, $x \in \Omega(\xi)$.

The problem of obtaining conditions for the Lagrangian, assuming existence of stationary trajectories, is a completely open question in the fractional framework. If the fractional derivative of the first order is considered, say α, equation (15.7) reduces to

$$\frac{\partial L(D^{(\alpha)}_{(\gamma)}q(\tau), q(\tau), \tau)}{\partial q} - D^{(\alpha)}_{-(\gamma)}\left(\frac{\partial L(D^{(\alpha)}_{(\gamma)}q(\tau), q(\tau), \tau)}{\partial \dot{q}}\right)$$

$$= \frac{1-\alpha}{1-\tau}\frac{\partial L(D^{(\alpha)}_{(\gamma)}q\tau), q(\tau), \tau)}{\partial \dot{q}}, \quad \forall \tau \in (\alpha, t), \tag{15.10}$$

where $D^{(\alpha)}_{(\gamma)} = D^{(\alpha)}_{(a_+)}$.

In the absence of fractional derivative, equation (15.8) reduces to

$$\frac{\partial L}{\partial x^k}(\dot{q}(\tau), q(\tau), \tau) - \frac{d}{d\tau}\left(\frac{\partial L}{\partial \xi^k}(\dot{q}(\tau), q(\tau), \tau)\right)$$

$$= \frac{1-\alpha}{1-\tau}\frac{\partial L}{\partial \xi^k}(\dot{q}(\tau), q(\tau), \tau). \tag{15.11}$$

Here $q: \ q(\tau, \xi)$, $0 \leq \tau \leq t$, $-\beta \leq \xi \leq \beta$ of class C^2 in (τ, ξ) from $A = q(0, \xi)$ to $B = q(t, \xi)$ in agreement with the given path $q = q(\tau)$ for $\xi = 0$, $L : \Re \times TM \to \Re$ is the Lagrangian, and $x = x(\tau, \xi)$ the coordinate point of $q(\tau, \xi)$.

Let us consider a useful example: a particle of mass m moving under the influence of a potential $U(q)$. The Lagrangian is given by $L = T - U$, where $T = (1/2)m\dot{q}^2$. In this case, the fractional Euler-Lagrange equations (15,11) easily yield

$$m\left(\ddot{q} + \frac{\alpha - 1}{\tau - t}\dot{q}\right) = -\frac{\partial U}{\partial q}, \tag{15.12}$$

representing the modified Newton's equation of motion, useful in our subsequent investigation.

15.3. Fractional covariant mechanics induced by modified scale laws

Let us assume that the dynamics of some point particle takes place on a fractal space-time manifold. If by $\vec{x}$ we denote the position vector of the particle, and by $\dot{x}^+_-$ the forward and backward mean velocities combined in terms of a unique complex velocity ([3], [4], [17]):

$$\frac{d}{dt}x(t) = \frac{1-i}{2}\frac{d^+x(t)}{dt} + \frac{1+i}{2}\frac{d_-x(t)}{dt} = \mathcal{V}$$

$$= \frac{v_+ + v_-}{2} - i\frac{v_+ - v_-}{2} \qquad (15.13)$$

then the Lagrangian function writes

$$L = L\left(x, \frac{(1-i)\dot{x}^+}{2}, \frac{(1+i)\dot{x}_-}{2}, t\right)$$

and the fractional action-like integral becomes

$$S[x](t) = \frac{1}{\Gamma(x)} \int_0^t L(x, \dot{x}^+, \dot{x}_-, t)(t-\tau)^{\alpha-1}d\tau. \qquad (15.14)$$

It can be shown that the corresponding fractional Euler-Lagrange equation is

$$\frac{\partial L(x, \dot{x}^+, \dot{x}_-, t)}{\partial x} - \frac{d}{d\tau}\left(\frac{\partial L(x, \dot{x}^+, \dot{x}_-, t)}{\partial \mathcal{V}}\right)$$

$$= \frac{1-\alpha}{\chi - \delta}\frac{\partial L(x, \dot{x}^+, \dot{x}_-, t)}{\partial \mathcal{V}}. \qquad (15.15)$$

It can also be proved that the complex time derivative operator takes the form

$$\frac{d}{d\tau} = \frac{\partial}{\partial\tau} + \mathcal{V} - i\mathcal{D}\Delta, \qquad (15.16)$$

where $\mathcal{D}$ is a parameter characterizing the fractal fluctuations.

Using (15.15) and (15.16), we can derive the modified Newton-Schrödinger equation. If Newtonian dynamics is considered, the Lagrangian of a closed system in the large scale domain is $L(x, \mathcal{V}, t) =$

$(1/2)m\mathcal{V}^2 - \Phi$, where ϕ is the associated potential energy. The fractional Newton's equation is then derived from (15.12) as follows

$$m\left(\frac{d}{d\tau}\mathcal{V} + \frac{\alpha - 1}{\tau - t}\mathcal{V}\right) = -\nabla\Phi, \tag{15.17}$$

In the absence of an external field, equation (15.17) yields

$$\frac{d}{d\tau}\mathcal{V} = \frac{1 - \alpha}{\tau - t}\mathcal{V}, \tag{15.18}$$

which is in contrast with the equivalence principle of Einstein's general relativity, leading to strong covariance principle, expressed by the fact that one may constantly find a coordinate system in which the space-time metric is locally Minkowskian. The complex momentum, in the fractional formalism, reads

$$P = m\left(\mathcal{V} + \frac{\alpha - 1}{\tau - t}x\right), \tag{15.19}$$

where the complex velocity is a function of the fractional action according to: $\mathcal{V} = \nabla S/m$.

15.4. Fractional Schrödinger equation and emergence of complex gravity

In order to derive the fractional Schrödinger equation, we introduce the complex wave function $\psi = \exp(iS/S_0)$, where the action S_0 has been introduced for dimensional reasons ([3], [4], [17]). Therefore $\mathcal{V} = -iS_0\nabla(\ln\psi)/m$. In view of (15.18) and (15.19) we then have

$$iS_0\left\{\left(\frac{\partial}{\partial\tau} + \frac{\alpha - 1}{\tau - t}\right)\nabla(\ln\psi)\right.$$

$$\left. -i\left[\frac{S_0}{m}\nabla(\ln\psi)\cdot\nabla((\nabla(\ln\psi)) + \mathcal{D}\Delta(\nabla(\ln\psi))\right]\right\} = \nabla\Phi. \tag{15.20}$$

Making use of relation

$$2\nabla(\ln\psi)\cdot\nabla(\nabla(\ln\psi)) + \Delta(\nabla(\ln\psi)) = \nabla\frac{\Delta\psi}{\psi} \tag{15.21}$$

and allowing $S_0 = 2m\mathcal{D}$, we find after a simple algebra

$$\frac{d}{d\tau}\mathcal{V} + \frac{\alpha - 1}{\tau - t}\mathcal{V} = -\frac{\nabla\Phi}{m}$$

$$= -2\mathcal{D}\nabla\left[i\left(\frac{\partial}{\partial\tau} + \frac{\alpha - 1}{\tau - t}\right)(\ln\psi) + \mathcal{D}\left(\frac{\Delta\psi}{\psi}\right)\right], \qquad (15.22)$$

and, by integration,

$$\mathcal{D}^2\Delta\psi + i\mathcal{D}\left(\frac{\partial}{\partial\tau} + \frac{\alpha - 1}{\tau - t}\right)\psi - \frac{\Phi}{2m}\psi = 0. \qquad (15.23)$$

One can define the fractional part of the potential energy as

$$\Phi_\alpha = 2im\mathcal{D}\frac{1 - \alpha}{t - \tau}, \qquad (15.24)$$

and hence the fractional Schrödinger equation still writes

$$\mathcal{D}^2\Delta\psi + i\mathcal{D}\frac{\partial\psi}{\partial\tau} - \frac{1}{2m}(\Phi + \Phi_\alpha)\psi = 0. \qquad (15.25)$$

It can be shown that the Schrödinger-Newton equation in curved and fractal space-time takes a similar form. In equation (15.25) the effect of the fractal formalism is expressed in terms of the complex potential Φ_α. Setting $M = im$, i.e. $M^2 < 0$ if $m > 0$, then Φ_α may correspond to the potential energy of the tachyons masses which decay in time. Moreover, by defining the quantity $\rho = \psi\psi^*$, the imaginary part of equation (15.25) writes

$$\frac{\partial\rho}{\partial\tau} + \text{div}(\rho\overline{V}) = \frac{1 - \alpha}{\tau - t}\rho, \qquad (15.26)$$

which can be recognized as the equation of continuity for a time-dependent (decaying) source term.

To illustrate this fractional formalism, let us now apply equation (15.25) to the problem of formation and evolution of the planetary system, where $\Phi = -GmM/r$, G being the gravitational coupling constant, $M = M_\odot$, and m the mass of the planet under consideration. In case of stationary dynamics with conservative energy, the energy operator is

$$2\mathcal{D}^2\Delta\psi + \left(\frac{E}{m} + \frac{MG}{r} - 2i\mathcal{D}\frac{1-\alpha}{T}\right)\psi = 0, \qquad (15.27)$$

where $T = \tau - t$. If we define $\mathcal{D} = GM/2\omega_0$, where $\omega_0 = 145 km/s$ is a universal constant having dimension of velocity [18], equation (15.27) can also be written as

$$2\mathcal{D}^2\Delta\psi + \left[\frac{E}{m} + \frac{M\overline{G}}{r}\right]\psi = 0, \qquad (15.28)$$

where we have assumed that $r = \omega_0 T$ and $\overline{G} = G + i(\alpha - 1)G$, *i.e.* G is the real part and $(\alpha - 1)G$ the imaginary part of the complexified gravity. Equation (15.28) is similar to Schrödinger equation for the Hydrogen atom with a complexified Bohr radius equal to $a_0 = 4\mathcal{D}^2/M\overline{G}$. Moreover, it can be shown that the ratio E/m of any planet is quantized and obeys the equation

$$\frac{E_n}{m_n} = -\frac{\overline{G}^2 M^2}{8\mathcal{D}^2 n^2} = -\frac{G^2 M^2(2-\alpha)}{8\mathcal{D}^2 n^2} - i\frac{G^2 M^2(\alpha-1)}{4\mathcal{D}^2 n^2}, \quad n = 1, 2, 3, \ldots$$
$$(15.29)$$

15.5. Scale relativity with fractional derivative for an arbitrary fractal dimension

In the light of results of fractional covariant mechanics induced by the modified scale laws, we presume again that the dynamics of a point particle takes place on a fractal space-time. To simplify the procedure, we shall use the fractional derivative with respect to space. In fact, the complex fractional derivative of order β is defined as

$$D_{(\gamma)}^{(\beta)} = \frac{1}{2}\left[D_{(b_+)}^{(\beta)} + i\gamma D_{(b_-)}^{(\beta)}\right], \quad 0 < \beta \leq 2$$

for space-like coordinates. If we denote by $\vec{x}$ the position vector of the particle, the forward and backward non-differentiable mean velocities are defined, respectively, by

$$\vec{v}^+ = \frac{d\vec{x}^+}{dt} = \frac{d\vec{X}^+}{dt} - \frac{d\vec{\xi}^+}{dt}; \tag{15.30}$$

$$\vec{v}_- = \frac{d\vec{x}_-}{dt} = \frac{d\vec{X}_-}{dt} - \frac{d\vec{\xi}_-}{dt}. \tag{15.31}$$

Here $d\vec{\xi}^+$ is the measure of non-differentiability, or, in other words, the fractal part of the displacement. As we can't favor $\vec{v}^+$ rather than $\vec{v}_-$, we define the fractional complex velocity by

$$\vec{V} = \frac{\vec{v}^+ + \vec{v}_-}{2} - i\frac{\vec{v}^+ - \vec{v}_-}{2}. \tag{15.32}$$

Denoting $d\vec{x}_\pm = d_\pm\vec{x}$, we still have

$$\vec{V} = \frac{\delta}{dt}\vec{x}, \tag{15.33}$$

where

$$\frac{\delta}{dt} = \frac{d^+ + d_-}{2dt} - i\frac{d^+ - d_-}{2dt} \tag{15.34}$$

is the complex differential operator in NSR. The real part in equation (15.32) corresponds to classical relativity, while the imaginary part corresponds to the non-differentiable factor. Since the position of the particle changes in time, we shall assume as a first approximation that the property $\langle d\vec{\xi}_-^+\rangle = 0$ (*i.e.* zero expectation of the fractal part of the space displacement) holds. Naturally, in the absence of fractional derivatives, the relation $d\vec{\xi}_-^{j,+} \propto dt^{1/D_f}$ is imposed. Carrying out a fractional derivative has the effect of replacing $D_f \to \overline{D} = \beta$, and therefore $d\vec{\xi}_-^{j;+} \propto dt^{1/\beta}$, $d\vec{\xi}_-^{j;+}d\vec{\xi}_-^{j;+} \propto dt^{2/\beta}$. and so on. In view of these notations, we can expand the total differential of the function $f(\vec{X},\tau)$ up to the third order term as

$$d_-^+ f = \frac{\partial f}{\partial t}dt + \nabla^{\beta/2} f \cdot \langle d_-^+\vec{X}\rangle + \frac{1}{2}(-\Delta f)_{ij}^{\beta/2}\langle d_-^+ X^i d_-^+ X^j\rangle$$

$$+ \frac{1}{3!}\nabla_i^{\beta/2} f(-\Delta f)_{ik}^{\beta/2}\langle d_-^+ X^i d_-^+ X^j d_-^+ X^k\rangle, \quad i = \overline{1,3} \tag{15.35}$$

where we have assumed that the mean values of the function $f(\vec{X},\tau)$ and its derivatives coincide with themselves. It is noteworthy the Riesz fractional potentials for the fractal space-time derivative defined as [19]

$$I_d^s f(x) = \frac{\Gamma((n-s)/2)}{\pi^{n/2} 2^s \Gamma(s/2)} \int_\Omega \frac{f(y)}{||y-x||^{n-s}}\, dy. \tag{15.36}$$

Here $s \neq n, n+2, n+4, \ldots$, which may be interpreted as negative powers of the Laplace operator with minus sign in $\Re^n : I^s f = (-\Delta)^{s/2} f$. It is not difficult to show that, for $\beta \to 2$, equation (15.35) reduces to

$$\frac{df}{dt} = \frac{\partial f}{\partial t} + \nabla f \frac{d_-^+ \vec{x}}{dt} + \frac{1}{2}\frac{\partial^2 f}{\partial x^i \partial x^j}\left[\frac{d_-^+ x^i}{dt}\frac{d_-^+ x^j}{dt} + \left\langle \frac{d_-^+ \xi^i}{dt}\frac{d_-^+ \xi^j}{dt}\right\rangle\right]$$

$$+ \frac{1}{3!}\frac{\partial^3 f}{\partial X^i \partial X^j \partial X^k}\left[\frac{d_-^+ x^i}{dt}\frac{d_-^+ x^j}{dt}\frac{d_-^+ x^k}{dt} + \left\langle \frac{d_-^+ \xi^i}{dt}\frac{d_-^+ \xi^j}{dt}\frac{d_-^+ \xi^k}{dt}\right\rangle\right]. \tag{15.37}$$

In view of (15.30) and (15.31), equation (15.35) can be written as follows;

$$d_-^+ f = \frac{\partial f}{\partial t}dt + \nabla^{\beta/2} f \cdot \langle d_-^+ \vec{X}\rangle + \frac{1}{2}(-\Delta f)_{ij}^{\beta/2}\left[d_-^+ X^i d_-^+ X^j\right.$$

$$+ \langle d_-^+ \xi^i d_-^+ \xi^j\rangle] + \frac{1}{3!}\nabla_i^{\beta/2} f(-\Delta f)_{jk}^{\beta/2}\left[d_-^+ X^i d_-^+ X^j d_-^+ X^k\right.$$

$$+ \langle d_-^+ \xi^i d_-^+ \xi^j d_-^+ \xi^k\rangle\Big], \; i = \overline{1,3}. \tag{15.38}$$

In equation (15.38), the following properties hold:

$$\langle d_-\vec{\xi}_-^{i;+} d_-\vec{\xi}_-^{j;+}\rangle = 2\mathcal{D}\delta^{ij}(dt)^{2/\beta}; \tag{15.39}$$

$$\langle d_-\vec{\xi}_-^{i;+} d_-\vec{\xi}_-^{j;+} d_-\vec{\xi}_-^{k;+}\rangle = \delta^{ijk}\frac{6\mathcal{D}^2}{C}dt^{3/\beta}, \tag{15.40}$$

where $\delta^{ij} = 1$ for $i = j$ and equals zero otherwise, $\delta^{ijk} = 1$ for $i = j = k$ and equals zero otherwise; quantity $\mathcal{D}^2/C$ is a constant of proportionality, augmented by the following relations:

$$\langle d\vec{\xi}_{-}^{i;+} d\vec{\xi}_{-}^{j;+} \rangle \gg 0 \qquad dt > 0; \tag{15.41}$$

$$\langle d\vec{\xi}_{-}^{i;+} d\vec{\xi}_{-}^{j;+} \rangle \gg 0 \qquad dt < 0; \tag{15.42}$$

$$\langle d\vec{\xi}_{-}^{i;+} d\vec{\xi}_{-}^{j;+} d\vec{\xi}_{-}^{k;+} \rangle \gg 0 \qquad dt > 0; \tag{15.43}$$

$$\langle d\vec{\xi}_{-}^{i;+} d\vec{\xi}_{-}^{j;+} d\vec{\xi}_{-}^{k;+} \rangle > 0 \qquad dt < 0. \tag{15.44}$$

Therefore, after dividing by dt and neglecting the terms containing differential factors, equation (15.38) can be written as

$$\frac{d}{dt} = \frac{\partial}{\partial t} + \nabla^{\beta/2} \cdot \vec{v} - \frac{i}{2}(-\Delta)^{\beta/2} \left[2\mathcal{D}(dt)^{(2/\beta)-1} \right]$$

$$+ \frac{1}{3!} \nabla^{\beta}(-\Delta)^{\beta/2} \left[\frac{6\mathcal{D}^2}{C} dt^{(3/\beta)-1} \right], \tag{15.45}$$

which is a fractional non-differential operator. The fractional geodesics equations are then

$$\frac{\partial \vec{V}}{\partial t} + \vec{V} \cdot \nabla^{\beta/2}\vec{V} - i\mathcal{D}(-\Delta\vec{V})^{\beta/2} \left[(dt)^{(2/\beta)-1} \right]$$

$$+ \frac{\mathcal{D}^2}{C} \nabla^{\beta}\vec{V}(-\Delta\vec{V}^{\beta/2} \left[dt^{(3/\beta)-1} \right]. \tag{15.46}$$

Equation (15.46) is identified as the fractional Burgers-Korteweg-de Vries equation in a fractal space-time. It reduces to a non-fractional equation for $\beta \to 2$. Researches expect many interesting consequences of equation (15.46), and investigations in this direction are promising.

References (Chapter 15)

[1] El-Nabulsi, R. A., *Fractional Nottale's scale relativity and emergence of complexified gravity*. Chaos, Solitons and Fractals, 42, 2009, p. 2924–2933.

[2] Mandelbrot, B., *The fractal geometry of nature* (Updated and augm. ed.), Freeman, New York, 1983.

[3] Nottale, L., *Fractal space-time and microphysics; towards a theory of scale relativity*, World Scientific, Singapore, 1993.

[4] Nottale, L., *Scale relativity and fractal space-time — a new approach to unifying relativity and quantum mechanics*, Imperial College Press, London, 2011.

[5] Kac, M., *The second Berkeley symposium on mathematical statistics and probability*, Berkeley, California, University of California Press, 1951.

[6] Laskin, N., *Fractional quantum mechanics*, Phys. Rev. E 2000; 62:3135.

[7] Laskin, N., *Fractional quantum mechanics and Levy path integrals*, Phys. Lett. A 2000; 268:298.

[8] Laskin, N., *Fractals and quantum mechanics*, Chaos, 2000; 10:780.

[9] Laskin, N., *Fractional Schrödinger equation*, Phys. Rev. E 2002; 66:056108.

[10] El-Nabulsi, R. A., *Fractional field theories from multi-dimensional functional variational problems*, Int. J. Geom. Meth. Mod. Phys 2008; 5(6):1.

[11] El-Nabulsi, R. A., *Fractional dynamics, fractional weak bosons masses, and physics beyond the standard model*, Chaos, Solitons and Fractals 2009; 41:2262-70.

[12] El-Nabulsi, R. A., Torres D. E. M., *Fractional action-like variational problems*, J. Math. Phys. 2008; 40:053521.

[13] El-Nabulsi, R. A., Torres D. E. M., *Necessary optimality condition for fractional action-like variational approach with Riemann-Liouville derivatives of order (alpha, beta)*, Math. Met. Appl. Sci. 2007; 30(15): 1931.

[14] El-Nabulsi, R. A., *Fractional Lagrangian formulation of General Relativity and emergence of complex spinorial and noncommutative gravity*, Int. J. Geom. Meth. Mod. Phys. 2009; 6(1):25.

[15] El-Nabulsi R. A., *Fractional action-like variational problems in holonomic, non-holonomic and semi-holonomic constrained and dissipative dynamical systems*, Chaos, Solitons and Fractals 2009; 42:52-61.

[16] El-Nabulsi R. A., *On the fractional minimal length Heisenberg-Weyl uncertainty relation from fractional Ricatti generalized momentum operator*, Chaos, Solitons and Fractals 2009; 42:84–8.

[17] Gottlieb I., Nica P., Agop M., *Scale relativity theory for an arbitrary fractal dimension*, Romanian Rep. Phys. 2008; 60(3):443.

[18] Nottale I., Schumacher G., Gay I., *Scale relativity and quantization of the solar system*, Astron. Astrophys. 1997; 332:1018.

[19] Chen W., Holm S., *Fractional Laplacian time-space models for linear lossy media exhibiting arbitrary frequency power-law dependency*, J. Acoust. Soc. Am. 2004; 115(4):1424.

Index